Psychology Revivals

Time, Space, and Number in Physics and Psychology

The crux of the debate between proponents of behavioral psychology and cognitive psychology focuses on the issue of *accessibility*. Cognitivists believe that mental mechanisms and processes are accessible, and that their inner workings can be inferred from experimental observations of behavior. Behaviorists, on the contrary, believe that mental processes and mechanisms are inaccessible, and that nothing important about them can be inferred from even the most cleverly designed empirical studies.

One argument that is repeatedly raised by cognitivists is that even though mental processes are not directly accessible, this should not be a barrier to unravelling the nature of the inner mental processes and mechanisms. Inference works for other sciences, such as physics, so why not psychology? If physics can work so successfully with their kind of inaccessibility to make enormous theoretical progress, then why not psychology?

As with most previous psychological debates, there is no "killer argument" that can provide an unambiguous resolution. In its absence, author William Uttal explores the differing properties of physical and psychological time, space, and mathematics before coming to the conclusion that there are major discrepancies between the properties of the respective subject matters that make the analogy of comparable inaccessibilities a false one.

This title was first published in 2008.

Time, Space, and Number in Physics and Psychology

William R. Uttal

Psychology Press
Taylor & Francis Group
LONDON AND NEW YORK

First published in 2008
by Sloan Publishing

This edition first published in 2015 by Psychology Press
27 Church Road, Hove, BN3 2FA

and by Psychology Press
711 Third Avenue, New York, NY 10017

Psychology Press is an imprint of the Taylor & Francis Group, an informa business

Publisher's Note
The publisher has gone to great lengths to ensure the quality of this reprint but points out that some imperfections in the original copies may be apparent.

Disclaimer
The publisher has made every effort to trace copyright holders and welcomes correspondence from those they have been unable to contact.

A Library of Congress record exists under ISBN: 9781597380157

ISBN: 978-1-138-83964-9 (hbk)
ISBN: 978-1-315-73329-6 (ebk)

Time, Space, and Number in Physics and Psychology

William R. Uttal
Arizona State University

2008
SLOAN PUBLISHING
Cornwall-on-Hudson, NY 12520
http://www.sloanpublishing.com

Library of Congress Control Number: 2007939517

Uttal, William R.
Time, Space, and Number in Physics and Psychology / William R. Uttal.
p. cm.
Includes bibliographical refrences and index.
ISBN 1-59738-015-6
ISBN-13: 978-159738015-7

Cover design by Amy Rosen

Sloan Publishing
220 Maple Road
Cornwall-on-Hudson, NY 12520

Printed in the United States of America

ISBN 1-59738-015-6

For Mitchan

Books by William R. Uttal

- Real Time Computers: Techniques and Applications in the Psychological Sciences
- Generative Computer-Assisted Instruction (With Miriam Rogers, Ramelle Hieronymus, and Timothy Pasich)
- Sensory Coding: Selected Readings (Editor)
- The Psychobiology of Sensory Coding
- Cellular Neurophysiology and Integration: An Interpretive Introduction
- An Autocorrelation Theory of Form Detection
- The Psychobiology of Mind
- A Taxonomy of Visual Processes
- Visual Form Detection in 3-Dimensional Space
- Foundations of Psychobiology (with Daniel N. Robinson)
- The Detection of Nonplanar Surfaces in Visual Space
- The Perception of Dotted Forms
- On Seeing Forms
- The Swimmer: An Integrated Computational Model of a Perceptual Motor System (with Gary Bradshaw, Sriram Dayanand, Robb Lovell, Thomas Shepherd, Ramakrishna Kakarala, Kurt Skifsted, and Greg Tupper)
- Toward a New Behaviorism: The Case Against Perceptual Reductionism
- Computational Modeling of Vision: The Role of Combination (with Ramakrishna Kakarala, Sriram Dayanand, Thompas Shepherd, Jaggi Kalki, Charles Lunskis Jr., and Ning Liu)
- The War Between Mentalism and Behaviorism: On the Accessibility of Mental Processses
- The New Phrenology: On the Localization of Cognitiive Processes in the Brain
- A Behaviorist Looks at Form Recognition
- Psychomyths: Sources of Artifacts and Mispresentations in Scientific Cognitive Neuroscience
- Dualism: The Original Sin of Cognitivism
- Neural Theories of Mind: Why the Mind-Brain Problem May Never Be Solved.
- Human Factors in the Courtroom: Mythology vesus Science
- The Immeasurable Mind: The Real Science of Psychology
- Time Space, and Number in Physics and Psychology

Contents

Preface

The crux of the debate between behaviorism and mentalist cognitivism focuses on the issue of accessibility. Cognitivists believe that mental mechanisms and processes are accessible, and that their inner workings can be inferred from experimental observations of behavior, or, to a lesser and more controversial extent, from introspection. Behaviorists, to the contrary, believe that mental processes and mechanisms are inaccessible, and that nothing important about them can be inferred from even the most cleverly designed empirical studies or insightful introspections. Behaviorists argue, therefore, that publicly observable behavior should be the central focus of psychology, whereas cognitivists contend that understanding the mind is an achievable goal.

It is clear that the conundrum of accessibility permeates much of modern psychological thinking. Nevertheless, the issue is rarely discussed overtly, and the controversy remains unresolved. Which side of the debate one falls on depends on some of the most fundamental and usually unspoken assumptions and prejudices of each individual psychological scientist.

One argument that is repeatedly raised by cognitivists and others is that even though mental processes are not directly accessible, this should not be an impenetrable barrier to unraveling the nature of the inner mental processes and mechanisms. Inference works for other sciences, so why not psychology? For example, the absence of direct accessibility, it has been pointed out by such scholars as the eminent physicist David Hestenes of Arizona State University, does not deter modern physical science from producing powerful theories of the nature of distant or microscopic matter that is equally invulnerable to direct examination. Great distances or ultramicroscopic size still leave traces of their physical properties in observations that are subject to measurement, and from those measurements the otherwise inaccessible properties can be inferred or derived. Thus, for example, it is possible to determine the age of, or distance to, a star by the pattern of electromagnetic radiation (e.g., the red shift) that does make its way to earth. Similarly, the invisible structure of an elementary physical particle can be deduced from the behavior of its constituent parts when they are ejected from a disintegrating atomic nucleus.

Cognitivists and other reductively oriented scientists then confront behaviorists with the challenge: If physics can work so successfully with their kind of inaccessibility to make such enormous theoretical progress, then why not psychology? Why then, this argument goes, should psychology be inhibited from the powerful tool of deductive inference any more than is physical science, just because we cannot directly examine its targets of inquiry? If this argument is correct, the door is open to a kind of cognitive reductionism in which the inaccessible mental processes and mechanisms can be analyzed and parsed into fundamental components or modules, causes identified, and interactions detailed. This is a powerful strategy that has served the physical sciences well. Behaviorism would then have to admit that it is only an incomplete, truncated approach to psychological knowledge and retire, if somewhat ungracefully in the style of most obsolescent psychological theories, from the field.

If the analogy between the properties of physical and psychological activities and dimensions is not, however, correct, then cognitivism would have to admit that it has set out on an impossible and intractable quest in its search for inner processes and mechanisms. One would imagine that its retreat would be equally ungraceful. Given the strong intellectual and emotional hold that either approach to psychological science has on its supporters, it is unlikely that a complete surrender of either school of thought will occur in the short run. Indeed, we might have to wait for a generational change, as we did during the last century of psychological history, for a comparable change in thinking.

The nature and comparability of physical and psychological inaccessibility, therefore, represent a formidable problem that has extreme importance in the development of psychological science. Indeed, this complex issue has troubled me for some years. In an earlier work (Uttal, 2007) I went so far as to express my doubts about this argument in the following way:

However, I must also admit that this is one of the most difficult challenges faced by any exclusively behaviorist approach, and I am not utterly convinced that physics and psychology can be compared in this regard. (p.83)

This concern set me off on a quest that culminates in this present book. The question asked here is: Are the properties of psychological and physical space and time sufficiently alike to require us to admit that inference works as well for psychology as for physics? The answer proposed here to this rhetorical question is that there are major discrepancies between the properties of the respective subject matters that make the analogy of comparable inaccessibilities a false one. An important corollary of this proposed answer is that the argument that the two fields are equally capable of overcoming their respective inaccessibility barriers is incorrect. As we see later, the main reason for this difference is that physical inference is supported by the general Cosmological Principle, which implies that the laws of the physical universe are the same everywhere, but psychology has no equivalent unifying principle.

The arguments that I present here to support these ideas are based on the differences between the dimensions of time, space, and number as they are observed in physics and psychology, respectively. The bases of these arguments are primarily empirical. We do know quite a bit about time and space in both psychological and physical phenomenology and are becoming increasingly aware of the differences between them.

Why, another version of the overarching question asks, has psychology been so recalcitrant to conventional mathematical analyses? Why should this science have been so incapable of being consolidated into broad, all-encompassing mathematical theories like those of Newton, Einstein, Bohr, and Heisenberg? Why, in other words, has psychology been so incapable of being fused into a pyramidal theory in which an increasingly small number of unifying concepts or laws describe an increasingly large number of empirical phenomena, observations, and findings? Instead, even its most ardent supporters agree that today's psychology is an aggregation of a large number of empirical findings that are, at best, described by fragmented and isolated microtheories that rarely speak to each other.

The general answers to which I have come to this set of questions is that psychological and physical space and time are not congruent. The ways in which space and time are dealt with in psychological phenomenology are quite different from their use in the physical sciences. Another reason is that behavioral responses are underdetermined. That is, they do not include enough information to provide a unique or robust answer to any questions about underlying mechanisms. An additional reason is that the properties of the powerful mathematical analyses that have served physics so well do not have the same power for psychological experiences. Psychology, instead, has turned from the deductive tradition of conventional mathematics to the stochastic and inductive methods of statistical analysis.

To understand these difficulties, we have to consider how the properties of mathematical interactions do or do not correspond to the properties of psychological phenomena. As it turns out, it seems that a plausible, even a compelling, argument can be made that conventional analytic mathematics exhibits fundamental properties and regularities that psychological phenomena do not. In other words, mathematics is built on certain rules of physical states that do not hold for cognitive states. Similarly, psychological space, time, and number do not follow the same laws as their physical equivalents.[1]

If this position is correct, it would explain why "mathematical psychology" is dominated by statistical and stochastic methods and ideas, rather than by the kind of conventional mathematics based on calculus and differential equations that has typically and historically been so productive for the physical sciences.

This brings us to the problems of predictability and extrapolation. The physical world permits deductive methods to predict the future to a remarkable degree for objects that range from the most microscopic to the most macroscopic. We have a remarkable record of success in computing such things as the trajectory of a missile or the day on which an eclipse will occur. There are debates between deterministic and probabilistic models of the physical world; however, there is no question that we do far better in predicting the behavior of a falling object than that of a thinking human. The question arising in this context is does this difference merely reflect a difference in respective variability or complexity, or does it reflect a real difference between physical forces and events and the causal factors motivating human behavior? In other words, are the differences between psychological and physical phenomena just quantitative, or are they fundamentally qualitative?

These questions establish the nature of the inquiry that occupies the remainder of this book. In one form or another, I repeatedly ask: Are the properties of the physical world and the methods that have been developed to represent them adequate to represent the properties of psychological function? In preview, I believe that the general answer to this question is that they are not. Therein lies the basic answer to the difficulties faced by psychology in developing comprehensive theory, as well as exhibiting the normal scientific trend toward ever more inclusive (pyramidal) and comprehensive theories. It explains why psychology remains, after centuries of concern, such a fragmented and much criticized activity by scientists in all fields—including its very own practitioners.

The differences between the properties of physical and psychological time, space, and number also explain why a true psychological science must attend to the aspects of behavior that are anchored to the physical world and eschew fantastic excursions into unsupportable theories and hypotheses concerning the nature of inaccessible cognitive processes. We do best when we work with well-defined physical stimuli and observed and measurable responses. We do least well when we attempt to infer cognitive or neural structures, mechanisms, and processes as explanations to account for the changes that occur between those stimuli and responses.

The concluding argument I wish to convey to my readers (at least to those who make it to the latter pages of this volume) is that the search for mentalist and cognitively reductive causes and explanations must be laid aside, as a child lays aside a beloved toy from its youth. Scientific psychology can only survive if it returns to its behaviorist roots and orientations. Only in this way can psychology become a full member of the scientific community.

This book presents one person's view of the reasons that psychology cannot be studied with the same tools as, for example, are mechanics and electromagnetics, and why, therefore, a molar behaviorism is the preferred future for our science. As with most previous psychological debates, such as the continuing one between behaviorism and mentalist cognitivism, there is no "killer argument" that can provide an unambiguous resolution. In its absence, an analysis of the differing properties of physical and psychological time, space, and mathematics may help to enlighten our thinking about why psychology is in the state it is.

ACKNOWLEDGMENTS

Although much of the work on this book has been done, as is usual in any writing project, in isolation, I have been greatly influenced in my thinking by a number of people. I have already mentioned my colleague Professor David Hestenes, whose thoughtful insight provided the initial stimulus for this book. David is a member of a small group of us who meet regularly to discuss different ideas, many of which are reflected in the present work. Our little Metaphysical Club also has included Peter Killeen, Warren Egmond, Michael McBeath, and other temporary members from time to time. They play an especially important part in my professional life since my retirement.

For the last four summers, I had enjoyed the joy and privileges of a visiting appointment at the University of Hawaii. Although a number of people made these visits possible, I have to single out Professor Patricia Couvillon, who gave new meaning to the Aloha Spirit. Pat greased the administrative skids and did everything possible to make these summers productive and pleasant. I am deeply appreciative of her help and that of my other colleagues who made these summers so wonderful.

I am also deeply indebted to Elizabeth Uhr who read the entire manuscript, helping greatly to improve style and structure.

Finally, but certainly not least, I am eternally grateful for the support of my dear wife, May, otherwise known as Mitchan, for a lifetime of love and wisdom.

NOTES

[1]A word of caution at the outset: In distinguishing between physical and psychological space and time or their respective laws, I am not proposing anything "supernatural" or even extraphysical. Brain states and their resultant mental activities are both the result of physical activity and of physical activity alone. However, my thesis is based on the foundation idea that because of the great complexity of the neural mechanisms that account for mind, there emerges a qualitative difference in their respective kinds of inaccessibility. This theme is developed further in later chapters of this book.

1

Time, Space, and Number in Physics

1.1 INTRODUCTION

One of the most compelling supportive arguments for a cognitivist approach to psychology is that psychology is not fundamentally different from physics. The mind is a natural process as are physical events. Both suffer from a kind of inaccessibility of the objects of their inquiry; physics because of the great distances to constellations and stars, the great masses of objects such as black holes and neutron stars, or the minute sizes of the subatomic particles (e.g., quarks, gluons, and hadrons) making up our universe. Psychology is equally constrained in its accessibility because of the failure of introspection and experimental assays to directly examine private, intrapersonal mental states. Furthermore, cognitive penetration[1] precludes even self-knowledge of one's own mental mechanisms and processes.

Because of the analogy drawn between inaccessibility in the two sciences, cognitivists suggest, there is no substance to the argument that mental responses are not equally amenable to the same kind of scientific inquiry that has both ennobled and enabled the wonderful accomplishments of the physical sciences. The analytic and inferential methods that have served physics so well can and should be applied to a psychological science of the mind. In this manner, cognitive psychology would be able to finesse the barriers to direct accessibility, and a mentalist psychology would be authenticated. Thus, our efforts to "scientifically" examine the inner processes and mechanisms of the mind that drive our behavior are justified, and the behaviorist arguments of an inaccessible mind overcome. Psychology, from this point of view, has nothing to apologize for; direct inaccessibility is a problem for all sciences.

The other side of the argument—the behaviorist position—asserts that although the issue may initially seem to be the same for psychology and physics, respectively, the natures of the respective inaccessibility barriers for each are quite different. That is, although direct accessibility is not possible for either of the two sciences, the nature of the differences between the two is much greater than they appear at first glance. The differences between the two views become obvious when one examines their relative complexity. The subject matters of physics are relatively simple, whereas psychology has problems of measurement and quantification that physics does not face. Physics has the enormous advantage of being anchored to the material world by a system of widely accepted dimensions and measurements that psychology does not enjoy. Thus, it is argued and the thesis of this book holds that the kinds of "inaccessibility" in the two domains are not comparable.

The strongest argument underlying the position that psychology operates under quite different and more severe constraints than does physical science is based on a premise that greatly advantages physics and does not work for psychology. That advantage is that the laws of physical nature that we can observe close at hand also seem to work every place else in the universe. Gravity may vary from place to place, but its laws are common here and at the ends (or, equally, at the beginning) of the universe. Electromagnetic waves travel at a constant speed in a vacuum regardless of their location. Although some (Aguirre and Tegmark, 2005) have speculated that the constants of nature that we measure on earth may differ in some other universes[2], there is still no evidence that suggests that physical principles, laws, and constants actually do vary from their earthly values within our universe. (For that matter, the idea of other universes is not universally accepted.) Thus, it is possible to safely assume that electromagnetism follows the same laws of propagation at the ends of our universe as it does here. Without this assumption, which underlies our interpretation of the significance of the ubiquitous red shift, virtually all of our physical theories would have to be modified and an entirely new conceptual model of the universe (as well as the most basic laws of physics) would have to be cast aside and replaced.[3] This presumption of the constancy of our physical laws, wherever they may be measured, permeates all of physical science from the macrocosm to the microcosm.

Unfortunately, an opposing view goes on to argue that psychology has not yet demonstrated an equivalent assumption of lawful constancy and simplicity, even at the relatively small human scale at which the mind-brain operates. Measurements of mental phenomena repeatedly show distortions of time and space that are not only nonveridical with measurements made with respect to the physical world, but also seem to be so irregular as to suggest that the mind may not be lawful in the manner that a science or a coherent mathematical theory requires.

The trophy that would be the end-product of this debate on the accessibility or inaccessibility of mental phenomena (if it could be resolved) is nothing less than the ultimate nature of psychological science. Should we accept the fact that inaccessibility is common to all sciences and pursue our goals of analysis of the mental processes and mechanisms in the manner championed by today's mentalist cognitive psychology? Or, to the contrary, should we accept the fact that mental inaccessibility is fundamentally different from physical inaccessibility, and that we are not likely to succeed in inferring, analyzing, explaining, or predicting the processes of the mind-brain with the same degree of success that physics has enjoyed? If the latter is the ultimate outcome of the debate, it is the case that a pure behaviorism would be the logical path to follow in the future development of a scientific psychology. Resolution of this debate is not yet at hand.

The purpose of this book is to review the scientific evidence that is germane to this great debate. My goals in this book are to:

1. Compare time, space, and number as they are conceptualized in physics and psychology, respectively. Both sciences have a long history of empirical studies of how these dimensions are processed in their respective domains. I search for order and, to the extent possible, draw synoptic conclusions that summarize what has been discovered.
2. Carefully examine how time and space are dealt with by mental processes and mechanisms and consider just how quantitative psychology can be. To do so requires that we examine the fundamental requirements for valid measurement in both physical and psychological processes and mechanisms.
3. Examine the properties of mathematics, originally evolved to model physical time and space, and their applicability to psychology. We cannot take for granted the assumption that the properties of physical time, space, and number are the same as those of mental activity and, thus, that the same kind of mathematics applies to both equally well. This discussion includes an examination of the reasons that statistics dominates psychology and conventional mathematical analysis dominates physics.

1.2 SOME TERMINOLOGY

In my earlier books, I struggled for many years in the hope that words such as mind, consciousness, sentience, qualia, perception, emotion, awareness, and a host of other mentalist terms could be defined in a way that would make possible a scientific (in the usual sense of the word) study of such private, intrapersonal responses as our thoughts. As time has gone by, I have realized that such a quest for precise definitions of mental phenomena and processes is (and has been for millennia) a waste of time. No definition of mind or any other mentalism has ever gone much beyond allegory, metaphor, or even worse, circularity.[4]

Over the years, although our usage of the terms *time* and *space* has been equally free and easy, like many of the basic mentalist vocabulary items, it is extremely difficult to establish a precise definition of what they mean. These difficulties are exacerbated when we extend the meaning to include the properties of psychological time and space. Indeed, my argument here is that the properties of physical time and space are considerably different from psychological space and time in their consensual meaning, quantifiability, and manner of measurement. I argue further that their respective vocabularies strongly support the argument that the two domains do not share equivalent constraints on the accessibility of their respective subject matters. Thus, the argument that physics and psychology share both a common problem (inaccessibility) and a common solution (formal or informal inference) for understanding their respective subject matters begins to unravel.

1.2.1 Time

Let's start with a preliminary effort to define *time*. As usual, the place to start, although with little expectation of achieving a satisfactory response, is a standard dictionary. My convenient computer dictionary provides a glut of different definitions but only the first four capture some of the meaning of *time* as a measurable dimension. Time is, according to this source:

1. A system of distinguishing events: A dimension that enables two identical events occurring at the same point in space to be distinguished, measured by the interval between the events.
2. A period with limits: A limited period during which an action, process, or condition exists or takes place.
3. A method of measuring intervals: A system for measuring intervals of time.
4. The minute or hour: The minute or hour as indicated by a clock.

(From *Encarta Dictionary of North American English* as embedded in Microsoft Word, 2003)

Obviously, none of these definitions provides the kind of precise denotation by reference to other well-established terms. However, the first comes closest to what is necessary for a technical definition and presages some of Albert Einstein's (1879–1955) early concern with non-simultaneity. That is, he inferred that some events that are simultaneous in one frame of reference are not necessarily simultaneous in another. Time, prior to the twentieth century, was considered to be a dimension marked off by events that flowed constantly and absolutely for all observers. Time as a limited period, as a moment in time, or as an interval has more limited connotations that differ from, but do not conflict with, the notion of a continuous dimension.

Some other quasi-technical definitions of time are:

- The continuum of experience in which events pass from the future through the present to the past.
- Fourth dimension: The fourth coordinate that is required (along with three spatial dimensions) to specify a physical event.

(wordnet.princeton.edu/perl/webwn)

A more complex definition of time has been offered by the anonymous author of the Wikipedia internet encyclopedia. Time, according to this author, is:

> A non-spatial linear continuum wherein events occur in an apparently irreversible order.

Although these definitions[5] introduce a number of new concepts, each is, on close inspection, as imprecise as are the dictionary definitions. First, definitions by exclusion (e.g., time is the non-spatial dimension) rarely help us to understand the nature of the referent. Second, some of the other ideas introduced here for the first time (e.g., linearity and irreversibility) are actually objects of extreme contention among philosophers and physicists as well as psychologists.[6] Third, some of the words (e.g., continuum and coordinate) used in these definitions are so obscure that they themselves require clarification and thus are of little help in defining *time*. In fact, some of these words might well be synonyms for *time*, and thus add nothing to our understanding at a technical level. This is a classic case of circularity in definition.

A more interesting and comprehensive definition of time has been provided by the Alliance for Telecommunications Industry Solutions (ATIS), an organization that obviously has a strong need for practical time measurements.

> **time: 1.** An epoch, *i.e.,* the designation of an instant on a selected, astronomical or atomic. It is used in the sense of time of day. **2.** On a time scale, the interval between two events, or the duration of an event. **3.** An apparently irreversible continuum of ordered events. **4.** That which characterizes, or is characterized by, the observed and apparently irreversible continuum of ordered events.
>
> (http://www.atis.org/tg2k/_time.html)

The fourth of these definitions is enticing, but it, too, is not exclusive. The phrase "continua of ordered events" is too inclusive and works for other dimensions (number sets) equally as well.

These definitions are probably as good as one can find. However, they are not much of an advance over Isaac Newton's (1643-1727) oft-quoted definition of time from the Scholium of Book III of his monumental *Principia* (Newton, 1687):

> Absolute, true, and mathematical time, of itself, and from its own nature, flows equably without relation to anything external, and by another name is called duration:[7]

One of the implications of Newton's definition of time is that time is uninfluenced by the contents or behavior of things embedded in it. As we see shortly, this is one of the major distinctions between his concept of time and that of modern special relativity theory. Furthermore, it might be well at this point to note in preview that subjective time hardly follows this maxim.

The most modern definition of time denotes it as the duration of a particular number of cycles of a particular kind of periodic oscillator. This definition ultra-precisely and quite effectively defines an interval along a time dimension in operational terms, but is equally ineffective in helping us converge on any better concept of time itself than do the more intuitive ones described above. At the present time, the standard oscillator for measuring time intervals is based on changes or transitions between atomic energy levels in an isotope of Cesium. Specifically:

> The second is the duration of 9,192,631,770 periods of the radiation corresponding to the transition between the two hyperfine levels [a spectral line] of the ground state of the Cesium-133 atom. (International Bureau of Weights and Measures: Paris)

This corresponds to an accuracy of "less than 2 parts in 10^{14}," the equivalent of a one-second drift in 1,400,000 years!

Of course, the original unit of time was not based on a Cesium isotope. Instead, it was originally defined as a fraction of other events, in particular a part of the day. One of the first pieces of evidence that physics and psychology deal with different concepts of time is that subjective time is not a close correlate of physical time. This was appreciated by Newton (as noted in Note 7) and is a major principle of modern psychological studies of time.

Even within the halls of physical science, time was not always dealt with in the same way. Classic Newtonian and Galilean time were absolute; clocks measured units of time that were independent of their motion through space. Nothing that could happen in space or on the earth could alter the absolute flow of time.

However, as scientific history has progressed, the very nature of our concept of time has changed. Definitions of time by modern relativistic physicists have become ever more impenetrable to the layperson. Time is no longer considered to be absolute but is defined in terms of the relative velocities of two frames of reference. The most concise expression of this lack of absolute time is to be found in the famous Lorentz equation:[8]

$$t' = \frac{t - \frac{v}{c^2} \cdot x}{\sqrt{1 - \frac{v^2}{c^2}}} \qquad \text{Equation 1.1}$$

where t' is the measured (or dilated) time at the velocity v, c is the velocity of light in a vacuum, and t is the time at a fixed or stationary place.

Time, according to this equation, changes as a function of the velocity of the observer. Equation 1.1 tells us that time appears to move slower in a moving object than to a stationary observer. (Because of the squared terms, this effect only becomes significant near the speed of light.) This slowing of time is a necessary outcome of the fact that, according to Einstein, the laws of nature are the same for all inertial frames of reference. It is also related to the fact that the mass of an object increases with velocity. The two combined factors imply that there must be an upper limit to velocities. Otherwise, infinitely large masses would occur in accord with the following expression.

$$m' = \frac{m^0}{\sqrt{\left(1 - \frac{v^2}{c^2}\right)}} \qquad \text{Equation 1.2}$$

where m' is is the relativistic mass at v and m^0 is the mass at $v = 0$. The upper limit on velocity was established by Einstein as the speed of light in a vacuum, that is, c.

1.2.2 Space

A similar expedition into the dictionary produces the following definitions of space:

1. The region beyond earth's atmosphere.
2. The region between all celestial bodies.
3. The three-dimensional expanse where matter exists.[9]

Obviously, the dictionary also fails us in this case. The meanings of space provided in 1 and 2 are specialized to completely different contexts. Only 3 begins to help, as it suggests that space is, or is defined as, a set of three dimensions. As we see later, however, it raises another contentious issue: Does space (and time as well) continue to exist in the absence of any matter or events? In other words, must space be occupied for its dimensions to have any meaning?

The answer to this question, like that of absolute time, has changed over the centuries. Classical, absolute space was also described by Newton in the Scholium of Book III of the *Principia* (Newton, 1687).

> Absolute space, in its own nature, without relation to anything external, remains always similar and immovable.

Other more conventional and intuitive definitions of space include:

> The infinite extension of the three-dimensional region in which all matter exists.

A somewhat more complex modern definition can be found in the online encyclopedia "Wikipedia":

> One view of space is that it is part of the fundamental structure of the universe, a set of dimensions in which objects are separated and located, have size and shape, and through which they can move.

As difficult as it is to define the older Newtonian idea of absolute physical space, definitions become increasingly more elusive when describing the relativistic space in which the dimensions and their units vary, depending on the relationship between a moving object and a stationary observer of that object's behavior. Like time, space also seems to be distorted by the velocity of an object relative to some fixed frame of reference. This is summed up in the Lorentz transformation for the spatial dimensions. For an object moving along one axis of space (say the x dimension), but with no motion along the other two axes, the distance $x'$is that the object appears to travel is represented by:

$$x = \frac{x - vt}{\sqrt{1 - \frac{v^2}{c^2}}} \qquad \text{Equation 1.3}$$

Comparable equations exist for the other two dimensions. Thus, according to this expression, distances is elongated as an object approaches relativistic velocities (i.e., as the object approaches the speed of light).

Furthermore, we now know that space is also distorted just by the presence of matter. Curvature of space occurs whenever massive objects are embedded in them, according to Einstein's general theory of relativity. Measurement of absolute space seems as elusive as absolute time at great speeds and for massive objects. Thus, the spatial and temporal properties of an object are not absolute in the modern sense; rather, they are dependent on its mass and velocity as well as its location. It was to unravel these perplexities that led to Einstein's earth-shaking contributions of special and general relativity.

On the human scale, there are also operational means of measuring or defining the units of the spatial dimensions. The standard for length, for example, for many years had been the distance between two marks on a platinum-iridium bar kept at the International Bureau of Weights and Measures in Paris. Nowadays, the standard meter is also defined by the temporal properties of a Cesium 133 clock. A meter is now considered to be the distance that light travels in a vacuum in a small fraction of a second, the small fraction being 1/299,792,458. This is a far more precise means of measuring distance than was possible with a metal bar standard. Of course, the original definition of a meter was based on something more down to earth—literally, the diameter of the earth itself. Specifically, the meter was originally defined as 1/10,000,000 of the distance from the Equator to the North Pole on a line that passed through Paris.

1.2.3 Space-time

The idea that space and time were independent dimensions began to be reconsidered in the late nineteenth century. Nikolay Ivanovich Lobachevsky (1792–1856) and Janos Bolyai (1802–1860) had both published papers on non-Euclidean geometries in 1829 and 1832, respectively. These geometries differed from the classic Euclidean geometries in certain of their postulates, the best known of which was their rejection of Euclid's concept of parallelicity. The important fact about these non-Euclidean geometries was next established by Eugenio Beltrami (1835–1900). In 1868, he wrote an important interpretation of the non-Euclidean geometries of Lobachevsky and Bolyai that showed that these geometries were more general and, thus, more powerful than the classic Euclidean representation of space. Indeed, the Euclidean geometries were actually included as special cases of the newer, non-Euclidean mathematics, just as the Newtonian time and space are special cases of relativity theory.

The fact that several different kinds of geometries could represent space equally well had an enormous impact on scientific thinking. Jules Henri Poincaré (1854–1912) realized that this fact suggested that no geometry was the "true" one, and that any geometry could represent space equally as well. They were all not only equivalent, but consistent with each other. Both Poincaré and Einstein realized that the implications of these mathematical ideas were so profound that they required an entirely new outlook on the physical world.

Einstein also appreciated something that becomes very important in the later discussion of the differences between physical and psychological time. He realized that Maxwell's laws of electromagnetism and the classic Newtonian concepts of space and time were incompatible with each other. One or the other had to be wrong; otherwise, the speed of light could be virtually unlimited, and the laws of the physical universe would permit infinite velocities, energies, and masses. To overcome this, Einstein suggested *that the laws of physics must be the same for all observers, regardless of where they are or at what velocity they are traveling*. This is the core idea of his world-shaking contribution of special relativity and has been formalized as a corollary of what is now known as the *Cosmological Principle*. One aspect of this principle means that the laws of physics are the same in one part of the universe as they are in any other part. Its importance for psychology lies in the fact that that there is no comparable principle uniting the mental and physical worlds. That is, the laws and properties of physical time and space are not the same as the properties of mental times and space. (I expand on this important concept later in this chapter.)

Both Poincaré and Einstein suggested that the old Newtonian and Galilean ideas of absolute time and space were no longer tenable in terms of this new context of non-Euclidean mathematics. Rather, based on challenges of how one went about measuring simultaneity and velocity, all of the dimensions and measurements had to be considered to be dependent on the state (i.e., velocity and location) of the observer and the frame of reference that an observer arbitrarily established in which to make his observations.

This arcane idea literally changed the universe. Time and space were no longer absolute; they were relative to the frame of reference of the observer! This is the heart of Einstein's Theory of Special Relativity. This theory also implies (as does the Cosmological Principle) that the laws of physics must be the same for any inertial frames of reference moving uniformly with respect to each other. If this were not true, then the speed of light would depend on where a light source was located and the direction from which and velocity with which the light was emitted.

The theories of relativity developed by Poincaré (1904) and Einstein (1905)[10] have led to many different reformulations of some of the basic ideas in modern physics. Masses, velocities, and distances, as already indicated, are no longer constants but vary with the state of the observer. Furthermore, as noted earlier, an important implication is that the velocity of light becomes an upper boundary on the velocity of any matter and possibly on the speed of transmission of information.

Furthermore, the very dimensions of space became variables depending on the frame of reference of the observer. The ether—a "fluid" in which the waves of light were supposed to be propagated—was no longer necessary and was promptly rejected by Einstein and the others. Most significant of all, however, was the new conceptualization of time as also dependent on velocity.

Poincaré (1902) summed up these developments as follows:

- There is no absolute space, and we only conceive relative motions.
- There is no absolute time. The equality of two durations can be defined only by convention, and we have no intuition of the simultaneity of events occurring in places. (quoted in Marchal, 2005, p. 4)

For Einstein, special relativity meant that there was no common clock time for different observers moving with different velocities. The Newtonian notion of absolute time had to be rejected in favor of an elastic time dimension.

There was one additional step in this development of modern ideas about time and space. Simply put, the "and" between time and space was subsequently deemed to be inappropriate. Shortly after the publication of Poincaré's (1904) and Einstein's (1905) work, Hermann Minkowski (1864–1909) suggested that space and time did not exist independently of each other. Rather, he proposed a geometry in which space and time were the indivisible parts of a single unified, inseparable, four-dimensional manifold designated as space-time. Mathematically, this new idea had advantages in simplifying the expression of the special relativity ideas expressed in Einstein's (1905) paper.

Einstein immediately understood the implications of Minkowski's concept. This unity of time and space was not just a mathematical fiction but a major reconceptualization of what we mean by time and space. Minkowski (1908/1952) captured the momentousness of the transition from time and space to space-time when he said:

> The views of space and time which I wish to lay before you have sprung from the soil of experimental physics, and therein lays their strength. They are radical. Henceforth, space by itself, and time by itself, are doomed to fade away into mere shadows, and only a kind of union of the two will preserve an independent reality. (p.75)

As poetic as these words may be, they still leave us with an intuitively inadequate "feel" for the meaning of time, space, or space-time. In fact, they exacerbate the difficulty of developing a definition—intuitive or denotative—of the four dimensions. The conventional view is that space is composed of, or can be resolved into, the three orthogonal dimensions—x, y, and z—and the one temporal one. Current theories meld these into a single entity known as space-time, but it still has a four-fold nature from the point of view of the human scale. The problem for physics is that time and space are no longer distinguishable; they have merged into a unified "thing"!

This is not the end of the story, however. New theories have been developed that seek to explain some of the contradictions and combine some of the theories by proposing something other than a four-dimensional universe. One modern conceptualization to which I previously alluded is *string theory*. String theory suggests that there are actually many dimensions of space-time that are not perceived by the human observer. It emerged as a result of the incompatibility of quantum theory (which did a fine job of describing the microuniverse) and relativity (which did a fine job of describing the macrouniverse). In an effort to combine the two into a unified theory, the concept of vibrating string-like dimensions was developed. As many as 26 or as few as 10 string types have been proposed in different theories, each of which represents a distinct type or dimension of space-time.

In recent years string theory has been challenged as being merely a mathematical abstraction that permits us to mathematically represent some conflicting ideas, but without meeting the usual standards of scientific verification (Smolin, 2006; Woit, 2006). According to these critics, many of its concepts have little relevance to physical reality. Others support it as the next great intellectual leap beyond relativity and do not demand intuitive appreciation of its various concepts. Some decades from now, the scientific community may resolve this controversy and perhaps further restructure our concept of space-time.

The important point of this extended discussion is that physical time and space are governed by a set of highly structured rules and theorems. Knowing the rules, it is possible to make precise measurements and predictions of the properties of objects that are not physically accessible to us. As this argument continues, the fact that psychological properties do not share this powerful constraint should become evident.

1.2.4 Is It Possible to Define Space and Time?

There is a possibility that the search for the definition of something as elusive as space or time or any combination thereof is going to be as difficult as the search for a precise definition of the mind. For example, the classic approach has been to consider space and time as a material framework within which objects can be located. This was a property of the absolute space of Newton's world view. As I noted earlier, Newton (1687), in his great *Principia*, proposed that time and space were absolute in that they existed without reference to any of their contents or events. However, there is an alternative way of thinking about these four dimensions. Perhaps time and space are only a set of measurements that take on meaning in terms of certain specific operations, such as measuring length or duration.From this point of view, any effort to define time other than operationally (for example, with a Cesium 133 oscillator) becomes, to say the least, very questionable.

The best and the brightest have had trouble defining space and time. Einstein (1917/2003) stated:

> It is not clear what is to be understood by "position" and "space." (p. 12)

This perplexity and his special intellectual struggle with the temporal concept of "simultaneity" were almost certainly among the intellectual keys that led to his proposing special relativity.

Some earlier philosophers (including Immanuel Kant, 1724–1804) also argued that there is no such thing as time. Rather, it is a hypothetical construct that we use to measure changes in our environment. Events occur in sequence, and things evolve and develop; we develop a metric and a rationale we call "time" for ordering these events. "Time" for Kant, was more a sensory experience, ineffable and immeasurable, as are any of the other mental states. Efforts to define it, therefore, are futile. Nevertheless, defining both space and time has been a goal of scientists for millennia.

A further complication is that the definition of space and time is constantly changing as new theories of the physical world unfold. Newtonian absoluteness has been replaced by relativistic ideas of space and time in which its measurements and units appear to differ depending on one's position and velocity. Thus, the old notions of time, space, mass, and velocity have been replaced by a new relativistic view in which these properties vary depending on conditions including relations and viewpoints. Absolute time and space disappear, and time and space as relative values take their place. Attempting definitions of these elusive rascals is "shooting at a moving target."

Whatever one's personal response to the changing interpretations, there are some constants about the way we define time and space. First, there is a general acceptance of the utility of absolute measures at the human scale. We all benefit enormously by being able to use units such as meters and seconds to determine positions and to spatially or temporally sequence things without having to worry about time being elastic or exhibiting unequal intervals. It is almost impossible to imagine how our world could survive if time and space were not dealt with as tangible, stable, and uncontroversial entities in an approximately absolute sense in our day-by-day activities. Relativistic expansions of time or elongations of space are still arcane matters of little interest to the general public. Relativistic transformations of space and time, as shown by Equations 1.1 and 1.2, become significant only for the very large and the very fast. On the human scale, we are concerned with the properties of something that is more akin to Newtonian absolute time and space. It is this standard against which we have to continue to compare the spatial and temporal attributes of mental phenomena until such time (and if) we develop a formal model that explains the contradictions, the subjectivity, and the paradoxical nature of psychological time and space.

When one seeks more rigorous definitions of the kind scientists use, the simple notions of time and space quickly become entangled in esoterica that require quite a bit more attention. At the very bottom of the problem, however, is the fact that time and space may not be definable. Definitions require that there be some more fundamental attributes or properties to which the definitions can be referred. Perhaps, as some scholars have suggested, time and space are simply the most fundamental and primitive attributes imaginable. In that case the search for definitions may be futile because they cannot be reduced to any more primitive concepts, and we are left only with the possibility of identifying their measurable properties or establishing useful units in some operational sense.

There has, indeed, been some progress in specifying the properties of ordinary time and space, even if, like the mind, their respective definitions may be elusive. It is on this edifice of progress that I now propose to base an additional argument that physical and psychological inaccessibilities are so different that we cannot use the utility of one to justify the other. In essence, I show that psychological space and time exhibit properties that are not those of physical time and space, and the mathematics that has evolved to meet the needs of the physical universe. If this argument is true, the unifying assertion and authenticating supposition of physics—the Cosmological Principle, which asserts that the laws of nature are the same everywhere—simply does not hold for mental processes. The laws of physics, classic or modern, do not work when we are dealing with the inaccessible nature of mind. In sum, albeit that inference of otherwise inaccessible processes is justified for physics, the parallel assumption does not hold for psychology.

1.3 THE BASIC PROPERTIES OF NUMBERS AND PHYSICAL TIME AND SPACE

To begin this analysis, I must first identify the properties of mathematics and of physical space and time that have been accepted by both Newtonian and relativistic physics. To begin, I examine some of the most basic notions of numbers and counting.

1.3.1 Cardinality

Although there are many technical definitions of cardinality in set theory, the one I wish to emphasize involves the idea of cardinal numbers. *Cardinal numbers* are those that are used to designate the quantity of a group of things. For example, numbers like 7 or 32 might designate the number of turkeys and ducks, respectively, in two flocks. Cardinal numbers are those associated directly with numerousness. There is no implication of order or interval in these numbers or even of the counting process that led to them, only that a match has been found between a symbol for a numerical value and the number of the birds in each flock. A cardinal number is equivalent to the last number obtained when the items in a group are counted, but it is not the same as the final count itself.

What cardinality does for us is to provide an initial basis for the representation of the property we call amount or numerousness—the quantity of whatever it is that is being measured. In physical terms, this may be the number of seconds, the number of kilograms associated with the weight of an object, or the distance in meters from one place to another. It is this basic concept of cardinal "amount," perhaps more than any other single concept in science, that allows us to develop more elaborate mathematical concepts.

In this context, it must be appreciated that cardinality is, therefore, the basic foundation of quantification and thus of the rest of the scientific enterprise. Without a basic concept of things differing in quantity, virtually all of the other dimensions and values we measure, as well as the process of measurement itself, would become essentially meaningless. The entire quest to quantify would be futile without an association between a cardinal number and the quantity or amount of things or stuff. Indeed, this primitive property is considered by some to be even more basic than the concepts of quality, time, or space. We can deal in the abstract with ill-defined kinds of entities or intangible concepts, but all of science becomes problematic if we do not have the basic concept of cardinality in hand. As we see later, this appreciation of cardinality is one of the first steps in the development of arithmetic skills in children. Indeed, it is often distinct from and usually occurs after children absorb the simple technical ability to count.

1.3.2 Ordinality

The basic concept of cardinality or quantity, however, is inadequate by itself to provide the basis for a system of measurement. It is also necessary to add to it a property designated as *ordinality*. By ordinality, I am referring to a property in which assertions are made about the relationships of objects along some dimension. The relational terms "more," "less," "bigger," "smaller," "earlier," and "later," and "further" or "closer" become meaningful only when we add the idea of order or ordinality to cardinality. When a dimension (in either time or space) is both cardinal and ordinal, we are not only saying that there are 7 turkeys and 32 ducks; we are also saying that there are *more* ducks than turkeys.

Ordinality plays a very important role in science; without it, none of our mathematical tools would work. Indeed, ordinality is a fundamental aspect of our ideas about causality. Physics, as I discuss later, depends on a particular order of events (the cause must precede the effect) in order to identify the forces involved in any process. The lack of ordinal relationships, the paradoxes of time that are so frequent in mental phenomena, is one of the most important reasons that the inaccessibility barrier is so impenetrable in psychology.

1.3.3 Monotonicity

Ordinality also implies *monotonicity*; that is, ordered values should steadily increase or decrease without any reversals that would permit a larger number to have a lesser value than a smaller one. Ten kilograms should always weigh more than eight kilograms if they are measured under identical conditions.

Although it is not necessary for a response dimension to exhibit equal intervals (we have many observations and measures that do not meet this criterion, one example being Weber's Law in psychology), a good measurement system benefits from equal intervals between its units. In general, physical time and space on the human scale are equal interval; a second occurring now has the same duration as one occurring later, and a meter in North America is the same as a meter in Australia (barring any relativistic effects, which I have already excluded from holding at the human scale).[11]

1.3.4 Continuity

It is popularly assumed that temporal and spatial dimensions are continuous. That is, they are composed of a continuous sequence of events or distances that cannot be subdivided into discrete and non-overlapping units however microscopic the cuts may be. At the practical human level, there appears to be a continuous flow of time and a continuity of distance along the spatial axes. Matter also seems to us to vary continuously.

There has, however, been considerable controversy concerning the appropriateness of the continuity assumption concerning ultramicroscopic values of time and space. The remarkable development of quantum physics in the twentieth century has led us to think about light and matter as being, in at least some of their fundamental manifestations, discrete. Photons and protons, electrons and even gravitons, as well as gluons and quarks, have been accepted in contemporary science as being ultimately small, discrete particles that represent the true discontinuous nature of ultramicroscopic physical reality. At this level, modern physics suggests that matter and energy are discrete.

If matter and energy are discrete and not continuous, then the question arises: Why not space-time itself? Indeed, a number of theoretical physicists, including Einstein himself,[12] suggested that, in fact, the four-dimensional universe is discrete. The idea evolved from the constraint on measurement known as the Heisenberg Uncertainty Principle. Werner Heisenberg (1901–1976) established that it was not possible to measure both the momentum and the position of an atomic particle at the same time (Heisenberg, 1927). Indeed, there was a trade-off; the more precise a measurement of position, the less precise the measurement of momentum. This equation is expressed as:

$$\Delta x \Delta p \geq h \qquad \text{Equation 1.4}$$

Where Δx is the uncertainty in position, Δp is the uncertainty in momentum, and h is Planck's constant (6.626068×10^{-34} m^2 kg/s).

One interpretation of this expression is that, since there is a limit on how finely space and time can be measured, there is a minimum size of the units of space and time. The current debate hinges on whether or not this limit is a property of the measuring instrument or represents a fundamental discreteness of space-time itself.

Mathematicians interested in number theory also have raised questions about the continuity of space. The problem arises because in number theory it can be proven that there is always a real number between any two real numbers, no matter how close they are to each other. This suggests that the dimension of real numbers may also be discrete since its "parts" can always be subdivided into ever smaller units of rational numbers (i.e., those that can be expressed as ratios of integers). Such a hypothesis of discrete space depends on how close the linkage is between the number system and space itself, a topic that winds off into philosophy as much as mathematics or physics.

Nevertheless, in the physical world at the human scale, the concept of continuous time and space still holds. I add it, therefore, to the list of fundamental properties of our physical time and space at our level. At cosmic levels, however, the problem becomes much more complex. New properties of space and time become salient. These new properties carry this discussion on to its ultimate—the Cosmological Principle—which plays such an important role in physics and whose absence portends so much trouble for psychology.

1.3.5 Isotropy

Modern relativity theory depends on some assumptions about the basic nature of the universe. These foundation assumptions are the end-product of a chain of logic originally based on astronomical observations that revolutionized thinking centuries ago. One of the most influential was the change from the earth-centered universe of Ptolemy (87–150) to the solar-centered one of Nicolaus Copernicus (1473–1543). This was an influential, even dangerous, concept not only because of the obvious fact that the earth no longer was the center of the solar system, but also because its implications stimulated one of the great paradigm shifts of thought. If the earth was not the center of the solar system or of the universe, was there any special privilege associated with our particular point of view? The Renaissance church was totally opposed to even considering a negative answer to this question but, as it turned out, the ultimately accepted answer laid the foundation of thinking that was to culminate in the relativistic theories of the twentieth century.

The ultimate answer to this rhetorical question was formulated in the now widely accepted principle of *universal isotropy*. The isotropic principle states that everything looks the same in any direction regardless of one's point of view. This was an enormous development. From some points of view, this questioning of a privileged point of view was not only the precursor of relativity theory, but also of a secular and equalitarian philosophy with all of its ramifications in religion, politics, and society.

Isotropy does not mean that every perspective is the same or that the same objects are present in all directions. Rather, it suggests that the nature of space is the same in all directions, even though the contents may differ in different directions. Thus, for example, if a force is applied to an object, the response is the same regardless of which direction the force is applied—all other things being equal. This concept is also known as Einstein's Equivalence Principle. It must be appreciated that isotropy is a statement about the nature of space, not about its contents.

At the human scale, the physical world is also functionally isotropic. The only modification of this generality is that there may be different forces operating on objects in local space. Gravity, for example, makes our world anisotropic; it is a highly directional (centripetal) force and establishes a privileged sense of up and down at the expense of left or forward. As we see later, this has a powerful effect on the perception and response of organisms to their environment. However strong this effect may be, it does not mitigate the theoretical assumption that space itself is assumed to be isotropic both at the cosmic and local scales.

In a mathematical context, isotropy means that there is an invariance of the orientation of the coordinate system being used. No matter how much you rotate or translate the axes used to describe a space, there should be an equivalence of the measurements being made. That is, measurements made in one direction should always be convertible to other directions, and forces in one direction should always have the same effect if redirected. In later chapters, we see that psychological phenomena do not appear to be cognitively isotropic any more than they exhibit other properties of the physical world. Our perceptual dimensions are not the same in every direction.

1.3.6 Anisotropy of Time

Unlike space, time is not generally assumed to be isotropic. This means that the dimension of time is not the same in every direction, as are the dimensions of space. Specifically, time progresses only from earlier events to later ones. It is impossible to run time backward, according to all except the most outlandish fringe theories seeking to justify some otherwise forlorn hope of time travel or some arcane theories of particle interactions.[13]

It is interesting to note, however, that there is no physical law or principle that says that time cannot run backward. Indeed, the concept that physical processes are the same regardless of the direction of time is a fundamental axiom of most of modern physics. This principle is referred to as *invariance* under time reversal.[14]

However, there are some practical constraints that suggest that time must be anisotropic and can run in only one direction. Collectively, these restraints have been referred to as "Time's Arrow" by such authors as Blum (1968) and Reichenbach (1956). The "arrow" has been interpreted in a number of ways, all of which are equivalent, interchangeable, and virtually synonymous.

1. On the whole, entropy or disorder increases in the universe. There is no way to go from disorder to order (i.e., you cannot unscramble an omelet), and thus time must run only forward.[15]
2. All real processes go toward a situation of greater probability, that is, equal energy (high entropy) distributions.
3. The increase in entropy is based on statistical considerations; it is unlikely that a random process with a large number of probabilities exactly reverses itself to its original state, even though it is likely that an ordered system becomes increasingly disordered by means of the same random processes.
4. The Second Law of Thermodynamics states that heat (energy) cannot flow from a cold object to a hot object. Thus, hot objects may heat cold objects, but not vice versa.

Thus, whatever the explanation and in spite of some fundamental controversies, physical time is generally considered to be anisotropic; that is, unidirectional, moving homogenously from the past to the present. As we also see later, psychological time seems not to always behave in the same way. Temporal anisotropy is challenged by memories, anticipations, and paradoxical distortions of physical time in our mental domain. Therefore, the assumption of the stability of laws of space and time used by physics to infer the nature of inaccessible objects and events from traces of their behavior may not hold for psychology.

1.3.7 Linear Causality

An important correlate of the anisotropic nature of time is that it is the past and the present that influence the future, and not vice versa. This property has been termed *linear causality*. It, along with the various formulations of Time's Arrow, adds to the general acceptance of an asymmetrical time that progresses monotonically from the past through the present to the future.

The problem of defining the causal relationship between events has, like so many other issues in science, engendered enormous historical controversy. How difficult it has been to define causation throughout history led Aristotle to propose that there were four different kinds of causes:

1. *The Material Cause*: The particular stuff or material out of which a thing is made.

2. *The Formal Cause*: The form or structure into which the material of a thing is made.
3. *The Efficient Cause*: The process or agent by which events occur, by means of which something happens, by which something is created.
4. *The Final Cause*: The goal or purpose for which something happens or something is made.[16]

As elegantly phrased as Aristotle was in expressing his four causes, it is only the third one that comes close to modern notions of causality. Hidden in the terminology of this kind of cause is the assumption that the cause always occurs before the effect. On the other hand, statements like the fourth of Aristotle's "causes" violate the near universal acceptance of the unidirectionality of time and lead to such inappropriate philosophies and theories as teleological evolution, religious ideas such as predestination or creationism, and such parapsychological nonsense as precognition—all of which imply that a cause may follow an effect, or that we can make a response to something that has not yet occurred.

One of the magnificent virtues of Darwinian evolution is that it was not goal-oriented, and that it is only the random events of the past and the present state that can affect or cause future developments. Darwinian evolution is not headed toward some goal (a retroactive cause) established by a creator; instead, at every moment of the present, the future is determined by the forces that are currently acting and those that acted in the past. The future of most complex systems is unpredictable for practical reasons: the small and numerous random influential causal events that occur over time. Species evolve because of the random selective forces of the past and the present; evolution is not driven by some future ideal goal, but by the need to survive at the present.

However strong these statements may seem, they are direct derivatives of the anisotropy of time and space. Biological processes, therefore, are directly in this line of modern physical thought and theory. Darwin's theory of evolution is, therefore, consistent with the anisotropy of time. Goal-oriented creationism is not a viable alternative theory of life; it violates some of the most basic laws of physics, including the unidirectionality of causal forces.

1.3.8 Homogeneity

Closely related to isotropy, and indeed assumed by some to be a derivative of it, is the concept of *homogeneity*. Homogeneity emerges from the twin facts that there are no preferred spatial directions and that there are no privileged locations—an idea whose roots are to be found, as noted earlier, in the Copernican revolution. The conclusion to which modern physicists have come from this premise is that every point in space is the same as every other one. As we see later, this property also does not apply to psychological processes.

Isotropy and homogeneity were among the assumptions that led to the development of what Einstein referred to as the Cosmological Principle, the idea that even though there may be local irregularities, if our sample is large enough, the universe is the same at all locations and in all directions. For Einstein, this Cosmological Principle had two parts:

1. The laws of physics are the same in all inertial frames.
2. The speed of light is a constant in all inertial frames.

A powerful implication of the Cosmological Principle is that the laws of physics work the same "there" as "here."

The first of these two laws is the glue that holds modern physics and cosmology together and permits inferences of the nature of directly inaccessible distant events. The second, although not directly applicable to psychological phenomena, is the basis for Einstein's famous $e = mc^2$ equation.

As the story I am telling unfolds, I argue that the laws of mental processing (i.e., the rules governing our perceptions and thoughts) are not equally constant at every point in psychological time or space. Most important of all, they regularly, if not always, conflict with the laws of physics as consolidated into the Cosmological Principle.

As a result, our mental life is distorted and altered by a host of what seem to be unlawful, irregular, anisotropic, and often illogical processes. The laws of physics are regularly violated in our perceptions and cognitions. Therefore, it has to be concluded that the facilitating assumption (for physics) that laws that work in one domain (the physical) cannot be assumed to hold in another (the psychological). The very important implication of this conclusion is that our efforts to infer the nature of the "mind" by drawing inferences from behavior are to no avail. The fundamental reason that this is the case and the basic explanation for the current state of psychological theory is that there is nothing comparable in psychology to the Cosmological Principle. Whereas physics can safely assume that distant and inaccessible objects and events follow the same laws as the more accessible ones, and therefore their nature can be inferred from what we observe, psychology cannot depend on this same saving assumption.

To sustain this conclusion, it is necessary to show that psychological time and space are not isotropic or homogeneous, nor are their properties always the same as the other properties of physical time and space discussed in this section. I show in subsequent chapters that there is ample evidence that this is the case. For this reason, psychology does not have the advantage enjoyed by physics in drawing inferences from its behavioral data about the nature of inaccessible processes and mechanisms.

1.4 THE PROPERTIES OF MATHEMATICS

The question now arising is: Can the fractionation and intransigence of psychology to be expressed in the form of a comprehensive formal theory be attributed at least in part to the fact that its fundamental properties and those of conventional mathematics are either in direct conflict or irrelevant to each other? In other words, are we trying to explain psychological properties that do not follow the rules of mathematical thinking that have served the physical sciences so well? An answer to this question may provide an understanding of the fragmentary nature of psychological science and its inability to converge on a unified theory in the form of a robust mathematical analysis of mental processes and mechanisms.[17]

Mathematics evolved over the years largely in response to the strong causal relationships observed in the physical sciences between material objects and forces. The physical world is driven by a relatively simple sequence of causal events (i.e., forces) that are well represented by mathematical terminology and a set of amazingly precise lawful relationships. Indeed, much of mathematics as we know it today has been derived specifically to meet the needs of the physical world.[18] The properties of physics and mathematics are congruent because observational physics has identified some of the properties of our world and mathematics has evolved to represent these properties.[19] The predominant reason that this approach has been successful is that there are regular, compatible relations between the process and mechanisms of the physical world and those of mathematics.

Whether a comparable regular relationship exists between psychological forces and mathematics is a matter of considerable concern to our science. R. Duncan Luce (1995) attacked this problem when he began his discussion of the rhetorical question, why should mathematics play a role in psychology? by expressing his conviction that:

> No one holds that all true statements we can make about a person's behavior are independent of each other. Some propositions follow as a consequence of others. (p. 2)

Thus, Luce suggested that mathematics has a role in psychology simply because there is sufficient causal and relational structure in psychological processes (comparable to the structure of the physical world) to answer his posed rhetorical question affirmatively. Unfortunately, Luce leaves many openings by limiting his assertion to "some propositions." The question remains open, therefore, concerning how big "some" has to be to justify the application of mathematics.

Luce was clearly aware that this general property of some kinds of interdependence might not be sufficient to justify a robust "mathematical psychology." He went on in this article to list a number of difficulties that mathematical psychologists faced when they attempted to apply mathematical methods to psychological processes. These included:

- "Each approach [to dealing with psychological variability] is to some degree unsatisfactory, and a fully satisfactory solution has not yet evolved." (p. 5)
- "All behavior obviously must arise from some internal activity. But it has been difficult to establish plausible connections between standard information processing ideas and some types of regular behavior." (p. 10)

- "A theory alleging only one mode of behavior may be easily rejected by a person having two or more available." (p. 12)
- "Little comparable invariance [compared to physics and genetic coding] has evolved in psychology. It is moderately rare to find a psychologist who, when confronted by a new set of data, invokes already known mechanisms with parameters estimated from different situations." (p. 13)
- "When each model is unique to a particular experimental situation, all of the model's free parameters must be estimated from the data being explained. Frequently, the resulting numbers of parameters outrun the degrees of freedom in the data. *This reflects a failure of the science to be cumulative, an unfortunate failure of psychology and social science that is widely criticized by natural scientists. I view it as one of the greatest weaknesses of modeling (and any other theory) in our science.*"(p. 13; italics added)
- "Coupling our lack of knowledge about local dynamic mechanisms with these statistical difficulties, it is hard to be optimistic about our ability to test these nonlinear models effectively." (p. 20)
- "I do not see a satisfactory solution for coping simultaneously with structure and error." (p. 22)[20]

The task ahead of us now is to see if, in the light of these potential problems, the modest criterion set by Luce (that is, the proposition that it is not true that "all true statements we can make about a person's behavior are independent of each other") is sufficient to link mathematics and psychology. The alternative, of course, is that this loose standard may be insufficient and, according to Luce, the whole idea of mathematical psychology "may not prove realizable in a deep sense; the attempt may prove to be a contradiction in terms, an oxymoron" (Luce, 1995, p. 2).

One way to attack this problem is to determine the properties of mathematics and see how well they match the properties of mental activity. Although, some of the properties of mathematics to which I now allude may seem complex and arcane, in fact the truly fundamental ones are relativity simple. Indeed, some of the most important are taught to us in our primary school days and are regularly used in our daily life. However familiar we are with these fundamental mathematical properties, all too often the discrepancies between mathematics as it is taught to us and psychological processes as we observe them are ignored. If it turns out that the properties of the two domains are not comparable or functionally equivalent, then we may make a terrible error in attempting to apply any kind of conventional mathematical methods to psychological processes. For example, one of the most basic properties of mathematics—additivity—is difficult to find in psychological experiments. The manner in which the wavelengths of the visual spectrum add is not modeled by simple additive relationships; qualitative changes often supplant quantitative predictions. One of the most obvious examples of this kind of unexpected perturbation is the shift of perceived color (a quality) with stimulus intensity—the Bezold-Brucke effect.

It is important to remember that most applications of mathematics to psychology are not based on the analytic forms used by physics in which causal relations are represented by deterministic, deductive methods. Instead, most mathematical psychological theories are statistical or stochastic in nature. Scientific psychologists typically do not deal with unique lawful relationships for single subjects, but rather invoke randomness and uncertainty to predict behavior for groups of subjects. In the present context, however, I am concentrating on the classical, deterministic form of mathematics that has graced and energized the physical sciences.[21]

My purpose now is to tabulate and to elaborate the properties of mathematics as we now know them. In subsequent chapters, we see how well or poorly psychological phenomena share these same properties.

1.4.1 Basic Arithmetic Properties

Arithmetic is the subfield of mathematics that deals with the basic operations of addition (+), subtraction (-), multiplication (×), and division (/)[22] and the rules by which they manipulate real numbers or algebraic expressions of real numbers. For a mathematical system to function reliably, these operations must follow certain rules. These proscriptions are known as the properties of real numbers and can be tabulated as follows:

1. *Commutativity*: The ability of equations to be true regardless of the order in which the components are arranged. There are two commutative laws, one for addition

$$A + B = B + A \qquad \text{Equation 1.5}$$

and one for multiplication

$$A \times B = B \times A \qquad \text{Equation 1.6}$$

respectively.

2. *Associativity*: The ability of equations to be true regardless of the order in which successive operations are carried out. Again, there are two associative laws, one for addition

$$A + (B + C) = (A + B) + C \qquad \text{Equation 1.7}$$

and one for multiplication

$$A \times (B \times C) = (A \times B) \times C \qquad \text{Equation 1.8}$$

respectively.

3. *Distributivity*: The ability of equations to be true regardless of whether the effects of a multiplier are carried out individually or collectively. For example,

$$A \times (B + C) = A \times B + A \times C \qquad \text{Equation 1.9}$$

4. *Equality*: Two results are equal if they have the same numerical value following the operations that produced them.

5. Identity: There are several identity operators that must hold for an arithmetic system for real numbers to be viable. (These are also known as the existence operators.) These include:

The additive identity property

$$A + 0 = A \qquad \text{Equation 1.10}$$

The multiplicative identity property

$$A \times 1 = A \qquad \text{Equation 1.11}$$

The Inverse Additive Property

$$A + (-A) = 0 \qquad \text{Equation 1.12}$$

6. *Meaningfulness*: Another property that is rarely included in this list is the property of meaningfulness. Suppes and Zinnes (1963) and Falmagne (2004) have pointed out the ratio of two temperatures as an example of a meaningless relation. The significance of the ratio (i.e., whether it is large or small) depends on the units, be they Centigrade, Fahrenheit, or Kelvin, as well as the absolute values of the measurements in whatever scale is used. On the other hand, the ratio of the *differences* between high and low temperatures on two days is the same regardless of the units used. (The size of the units essentially cancels out in this case, but not in the first.) The point is that the numbers that are used to represent variables are meaningful only if they do not depend on the choice of units. If we are to assign a number to some process or mechanism, we must be assured that it is not a fiction produced by some inadvertent misuse of the arithmetic operations. For psychologists, this is a particularly important problem since so many of the processes measured are little more than reified extrapolations of experimental findings and actually may not be meaningful in the sense used here. Furthermore, the scales of the dimensions of psychological measurement are usually obscure.[23]

7. *Spatial and Temporal Properties*: Finally, it is worthwhile to repeat that in order to be useful to physics, the arithmetic operators must operate on spatial and temporal dimensions that are characterized by the known properties of physical space and time. In order for arithmetic, much less advanced mathe-

matical procedures, to be applied, time must be homogeneous but not necessarily isotropic. Space, on the other hand, must be homogenous and isotropic. Both must be continuous and monotonic. Most particularly, causal forces must exert their effect from the past towards the present and the future. As we also see later, these properties do not necessarily adhere to psychological processes and mechanisms.

1.5 ON PSYCHOLOGICAL QUANTIFIABILITY AND MEASUREMENT

The field of mathematical psychology is based on two fundamental assumptions. One is Luce's (1995) suggestion that at least some of "the true statements that we can make about a person's behavior" are dependent on each other. Thus, he concluded that "some propositions surely follow as a consequence of others" (p. 2). This assumption may be summarized as the Principle of Causal Relationship. It is not only a principle of mathematical psychology, but also of any mathematical scheme that purports to derive theorems from axiomatic foundations in an orderly way. It says that events follow from each other in a manner that indicates some sort of functional causation, implication, or even equality.

The second fundamental assumption of mathematical psychology is that psychological parameters and dimensions are actually quantifiable. This assumption may be summarized as the Principle of Psychological Quantifiability. This principle has rarely been made explicit by experimenters before they conduct their experiments. Unfortunately, the necessary conditions for quantification and measurement are not always present when studying mental processes.

The advantage of any successful system operating under these two assumptions is that powerful and precise mathematical methods can be applied in a way that permits us to transcend the weak and imprecise verbal descriptions of psychological functions. All too often in psychology, underdetermined hand waving is presented as a theory, description, or explanation of some psychological phenomenon. Regardless of what form of mathematics (if any) may ultimately turn out to be appropriate for the representation of a particular kind of psychological activity, all such methods require that there be both some "causal" or "consequential" relationship between the events of behavior, and that the processes under study be quantifiable.

Quantifiability, in particular, is a sine qua non for the development of any mathematics-based science. As we saw in earlier sections of this chapter, all of physical science depends on certain properties of the quantifiability of time, space, and numerousness to provide the coherence and regularity necessary for the application of not only mathematics, but also any systematic method of study (e.g., taxonomic classification) to a domain of knowledge. The dimensions of time and space must be measurable in units that are dependable and stable from situation to situation, and that follow the laws of arithmetic manipulation. Unless the dimensions and units of psychological mechanisms and processes are also dependable, then formal models of an axiomatic-deductive nature would be impossible. If 1 + 1 does not always equal 2, trouble is ahead for any putative science.

Clyde Coombs (1950), the eminent University of Michigan mathematical psychologist and a highly respected friend and colleague from my Michigan days, put it in the following words:

> The concept of measurement has generally meant the assignment of numbers to objects with the condition that these numbers must obey the rules of arithmetic. This concept of measurement requires a ratio scale—one with a non-arbitrary origin of zero and a constant unit of measurement. (p. 145)[24]

It is important to appreciate that the requirement for a non-arbitrary zero also implies that the metric being used is non-arbitrary. (A *metric* in a formal sense is a "geometric function that describes the distances between pairs of points in a space.") For physics, we have a very clear idea of what the metrics are. However, for most cognitive, mental, or psychological spaces, the metric is often arbitrary and inadequately anchored to any physical dimension, if not just downright mysterious. In many cases, zero points are arbitrary. Measurement in psychology is hampered by this arbitrariness. Efforts to make psychological metrics more robust are ongoing, but generally unsuccessful.

I continue this discussion by now considering in greater detail some of the properties that make measurement possible. One of the most significant is, as Coombs proposes, that it is necessary for the numbers to follow a "ratio scale." Ratio scales as defined by Stevens (1951) are:

> those most commonly encountered in physics, and they are possible only when there exist operations for determining all four relations: equality, rank order, equality of intervals, and equality of ratios. (p.28)

Another property that is characteristic of a ratio scale is the idea that has been called "meaningfulness" by Stevens (1946), Suppes and Zinnes (1963), and Falmagne (2004) and which was introduced on page 24. Meaningfulness requires that measurements "be invariant with respect to changes in the units of its variables" (Falmagne, 2004, p. 1342).

Quantitative measurement, therefore, is not something that can be applied to any observation a priori without serious and detailed examination to determine if the necessary conditions are satisfied and the properties present. To emphasize this point, we must consider what both Stevens (1951) and Siegel (1957) had to say about different kinds of measurement. Both identified four different ways in which "measurement is understood to mean the process of assigning symbols to observations in a consistent manner" (Siegel, 1957, p. 15).

The four-fold system they use to define valid "measurements" made under this very loose definition includes the following four categories arranged from the weakest form of measurement to the strongest:

- Nominal or classificatory scales in which any kind of a symbol is attached to an item. The only property of this kind of measurement is that different items with the same name are "equivalent" or "equal" to each other.
- Ordinal or ranking scales in which the concepts of "greater than" or "less than" are added to the concept of nominal equivalence—"equal to."
- Interval scales in which we add the concept of equal intervals to those "equal to," "greater than," and "less than" criteria. However, no fixed zero is specified and, being undefined, it is arbitrary.
- Finally, ratio scales add the concept of equal ratios and a true (i.e., non-arbitrary) zero to those concepts defining an interval scale. Equal ratios imply that the values of the ratio do not change when the units change. It is this kind of scale that Michell (1999) and Coombs (1950) defined as the sine qua non of quantification and thus the basis of valid measurement.

Siegel (1957) vigorously made the point that only certain kinds of manipulations are possible for each of these four levels. Table 1.1 tabulates what he believed to be the statistical methods suitable for each level.

The question now arising is—do psychological phenomena exhibit the properties of meaningfulness and ratio scales required by Coombs', Siegel's, Michell's, and Stevens' definitions so we can consider them to be quantifiable and thus measurable?

TABLE 1.1
Levels of Measurement and Appropriate Statistics

Scale	*Appropriate Statistics*
Nominal	Mode
	Frequency
	Contingency coefficient
Ordinal	Median
	Percentile
	Spearman r
	Kendall τ
	Kendall W
Interval	Mean
	Standard Deviation
	Pearson Product-Moment Correlation
Ratio	Geometric Mean
	Coefficient of Variation

Source: S. Siegel, Nonparametric Statistics (*The American Statistician*, 1957).

To summarize, here are the properties that have emerged in the discussion so far. To be quantifiable and measurable, psychological phenomena must:

- follow the laws of arithmetic, especially additivity
- possess a non-arbitrary zero and a non-arbitrary metric
- possess a constant unit of measurement
- exhibit equality
- exhibit rank order, specifically ordinality. (ordinality implies monotonicity, continuity, and causality)
- exhibit equality of intervals
- exhibit equality of ratios
- be capable of being transformed from one system to another simply by multiplying each value in one system by a single number.

Although this list may seem overwhelming and too restrictive, it must be appreciated that it is because the phenomena meet all of these properties that mathematics and formal physical theory are so successful. In the absence, for example, of equal unit intervals, the operations and measures that are used to define the units of time would be meaningless. Time without equal intervals would be characterized by units that varied capriciously depending on the task. One could never know when something happened and if the necessary properties of equality or meaningfulness existed.

Time as measured in the psychological laboratory, however, does not meet all of these criteria. As only one example, subjective time seems not to exhibit the equal-interval property phenomenologically. Does this begin to suggest that time perception is not quantifiable? Does this begin to suggest that mathematical psychology in this context is an example of Luce's hypothetical "oxymorons"? It is becoming increasingly clear that the answers to these rhetorical questions are "yes" and "yes," respectively.

Efforts to define the necessary conditions for quantifiability have been longstanding. A classic formulation consisting of seven rules was proposed by Holder (1901), but a modern reformulation into five more concise rules has been proposed by Michell (1997). Some of these rules are the usual arithmetic ones, but others add specificity to a putative definition of quantification. Michell listed these conditions for quantification as follows:

1. Any two magnitudes of the same quantity are either identical or different and, if the latter, there must exist a third magnitude, the difference between them—i.e., for any a and b in Q, one and only one of the following is true:
   - (i) a = b
   - (ii) there exists c in Q such that a = b + c
   - (iii) There exists c in Q such that b = a + c

2. A magnitude entirely composed of two discrete parts is the same regardless of the order of composition—i.e., for any a and b in Q,

$$a + b = b + a$$

[This is the commutative law of arithmetic already mentioned.]

3. A magnitude that is the part of another magnitude is also the part of the same magnitude, the latter relation being unaffected in any way by the former—i.e., for any a, b, and c in Q,

$$a + (b + c) = (a + b) + c$$

[This is the associative law of arithmetic previously mentioned.]

4. For each pair of different magnitudes of the same quantity there exists another between them—i.e., for any a and b in Q, such that $a > b$ there exists c in Q, such that $a > c > b$.

5. Given any two sets of magnitudes, an upper set and a lower set, such that each magnitude belongs to either set but none to both, and each magnitude in the upper set is greater than any in the lower, there must exist a magnitude no greater than any in the upper set and no less than any in the lower—i.e., every non-empty subset of Q that has an upper bound has a least upper bound. (p. 357)

Michell goes on to note that conditions 4 and 5 in this list "ensure the density and continuity, respectively, of the quantity" (p. 357), thus linking these criteria for quantifiability with the properties of physical space and time as previously described.

The important thing about all of these "conditions," "criteria," or "principles" of what constitutes a quantitative dimension is that quantifiability is not simply the application of numbers or even the operation of measuring something. Stevens' (1951) suggestion that "measurement is the assignment of numerals to objects or events according to rules" ignores the problem of whether or not the objects or events are susceptible to the assignment of numerals. Instead, three points should now be clear:

- To be quantifiable in a strict sense, the dimensions being measured must exhibit certain properties and interrelationships.
- Not all dimensions meet these qualifications.
- Psychological mechanisms and processes should be critically examined to see if they exhibit these properties prior to efforts to measure them.

Unfortunately, the last of these three admonitions is almost never honored in psychological research laboratories.

1.6 INTERIM SUMMARY AND A PREVIEW

The essence of the question I ask in this book can be formulated in several ways.

- Do psychological properties exhibit the properties of measurement and quantification collectively and individually in a way that permits us to measure, interpret, and infer what are otherwise inaccessible events?
- Do psychology and physics differ so greatly that what can be done in physics cannot always be done in psychology?
- Do the properties of psychological dimensions, events, and responses permit us to apply psychophysical methods (e.g., the method of magnitude estimates) as well as other essentially mathematical procedures such as Multidimensional Scaling and Factor Analysis with any assurance that the obtained categories are meaningful?

Although some of the answers to questions like these that would separate psychology from other sciences are likely to be repugnant to psychologists, there has to be some modicum of concern with such fundamental conceptual issues if psychology is to flourish. However, the amount of concern attributable to these issues has been insufficient. Michell (1997) and others have suggested that while psychologists have spent an enormous amount of time and effort developing methods for making quantitative estimates, they have spent very little time asking the most basic questions about fundamentals, such as the quantifiability and thus the measurability of psychological observations. Speaking of the problems of measurement, Michell noted that psychologists:

> have adopted their own, special definition of measurement, one that deflects attention away from the scientific task [of showing that the relevant attribute is quantitative] ... From Fechner onwards, the dominant tradition in quantitative psychology ignored this task. (p. 355)

A major purpose of this book is to join the few others who have considered this issue. I propose to do this by examining a sample of the experimental literature and to compare what appear to be the properties of psychological dimensions with the conditions for quantifiability listed here.

Cosmological and subatomic physics have transcended the challenge to accessibility imposed by huge distances and microscopic scales by depending on a system of ideas elegantly summed up as the Cosmological Principle. The basic idea is that the world is pretty much the same everyplace, and that the laws of physics work everywhere the same. Therefore, any observations that are made of the very distant or the very small can be interpreted in the same manner and by the same (or generalizations of) laws as local observations.[25]

A major thesis of this book is that there is nothing that corresponds to the Cosmological Principle in psychology. This powerful tool permits us to overcome the practical problems of inaccessibility due to distance or scale in physics and allows us to draw powerful and reasonable inferences about the nature of far away or very small places. Although the size and time scale of mental processes is the same as local physical processes, there is no way to assure a priori that the laws and properties of mental activity are the same as those of the material world of which they are a part. Indeed, as the remaining chapters of this book show, there is ample evidence that psychological processes and mechanisms are characterized by properties and relationships that differ greatly from those of the physical environment. Thus, the validating anchor that physics enjoys (laws and properties "here" are the same as those "there") may not hold for psychology.

To the contrary, psychological processes and mechanisms appear to operate by irregular and paradoxical laws and to exhibit properties that vary greatly from those of physics and mathematics. This may mean that measurements appropriately made of behavioral responses cannot be depended on to be accurate representations or even spatially or temporally congruent with what is actually going on in the mind. Instead, psychological time and space operate along dimensions that do not exhibit the same properties of, for example, homogeneity, isotropy, or even the flow of causation from the past to the present, as does the physical domain. This fact has a number of implications:

1. Conventional deductive mathematics, with its orderly step-by-step derivations from axioms to theorems, may not apply to psychological functions. The fact that most theories of psychological function are based on stochastic (probabilistic) rather than deterministic methods is a covert expression of the inappropriateness of conventional mathematics for psychology. Statistics, it may be argued, finesses the possibility of direct causation by substituting correlations—correlations that are highly underdetermined regarding the forces at work. Thus, like many problems scattered throughout the scientific world, but to an exceptional degree in psychology, questions about mental activity are often "ill posed," in that the ensemble of behavioral observations may not contain adequate information to answer the question.
2. If this suggestion is correct, the inferential linkage between observations of behavior and otherwise inaccessible cognitive processes is broken. This does not mean that behavioral observations are not quantifiable and subject to description and analysis by conventional mathematics or statistics or both. Behavioral observations do have the properties and meet the conditions for quantifiability and analysis; it is the inferences, theories, and explanations drawn from them that are not only problematic but in a formal sense intractable!
3. The ubiquitous nonveridicality between stimuli and cognitive responses (as reflected in behavioral observations discussed in Chapters 2 and 3) suggests that psychological transformations are being driven by causes, properties, and operations that differ significantly from the causes, properties, and operations of the physical world and the most powerful tool in its repertoire—mathematics.

How these properties, laws, and operations differ between the two domains is the major issue in understanding why psychological theory has been so perpetually disappointing. There can be little argument that each so-called "psychological theory" is constrained to severely limited problem areas; it has not pyramided into anything approximating a global theory. This also explains why it has proven to be difficult to discriminate one "plausible" theory from another. To put it succinctly, behavior is underdetermined.

Physical space and time define the limits of the world of life and sentience in which we exist, just as they do that of inanimate objects. Although there is plenty of controversy and debate concerning the exact nature of the properties of space and time, important developments have occurred in the past century that have revolutionized our concepts of their nature. Newtonian ideas described a simple world of human scale dimensions and relationships. Relativistic physics, by means of such global criteria as the Cosmological Principle, facilitated understanding of the nature of these dimensions. However, relativity theory and quantum mechanics did not "overturn" or topple Newton's ideas of time and space. Rather, they showed that there were implications that went beyond the human scale and our ability to directly observe it. Newtonian physics, however incomplete, was shown to be contained within the broader relativistic and quantum theories. That is, the idea of the absolute world worked except under the most extreme conditions of motion and scale.

Physical time and space (or space-time if you prefer) exhibit the properties listed and described in this chapter. The task at hand is to compare the properties of psychological time and space with those of physics. By doing so, we can build a realistic appreciation of the extent to which mental events can be measured and the role that mathematics can play in our conceptualization, description, and explanation of psychological mechanisms and processes.

The next important point that must be made explicit is that even the psychological world is, at some level, based on these same properties of physics. Materialism is the prime postulate and the most fundamental consideration in the development of any scientific world view. Any alternate hypothesis that invokes non-physical dimensions and supernatural attributes or that proposes processes and mechanisms that conflict with the physical world violates this prime postulate and must be rejected a priori. Thus, however irregular behavioral and cognitive processes may be, they are still a result of the operation of a material world, specifically the nervous system. Their irregularity, the contradiction to the Cosmological Principle, and the absence of any corresponding facilitating principle results from complexity and adaptivity, not from any supernatural or mystical forces. One implication of the embedding of even the strangest mental phenomenon in the material world is that no psychological theory that violates the laws of physics is acceptable. Hypotheses such as parapsychological processes, mental life after physical death, and efficacious prayer all invoke extraphysical forces and have to be excluded a priori from the scientific enterprise.

I appreciate that this is not the current state of thinking in most of the world, nor has it been throughout history. Many non-physical and supernatural ideas permeate society, and the point of view expressed here is, without question, a minority view. Unfortunately, any reliance on the non-physical or the supernatural inevitably leads to some gross misunderstanding and misinterpretation about some of our scientific activities, especially those associated with the interpretation of the accessibility or inaccessibility of mental matters. Some may attempt to use some of the arguments presented in this book for a non-quantifiable, inaccessible, or immeasurable mental world as a basis for an attack on the fundamental physicalism, materialism, or monism that must be the logical foundation of any science. This would be a gross misuse and misunderstanding of the thesis presented here. Complexity and immeasurability are not arguments for dualism!

What I intend to show in this book is the nonveridicality of some of the most basic properties of the physical world and some of those of the psychological world. It is on this foundation that I base my argument that physical inaccessibility (of distant galaxies, for example) is fundamentally different from mental inaccessibility (of our decision processes, for example). In particular, I show that the properties of cognition differ from some of the properties of the material world, and that, as a result, the lawful, causal relations that exist in physics may differ from those that characterize psychological processes and mechanisms. This means that the basic properties of physical time, space, and mathematics cannot be assumed to hold for the mind as they do for cosmology. Another set of properties must be invoked, a set that requires a different kind of approach that may not be explanatory in the same manner as the physical world.

Unfortunately, like so many other problems that occur at the boundary between science and philosophy, there are many secondary issues that have to be resolved to make the case that psychology, not being congruent with physical dimensions and laws, suffers a greater disability than does physics in dealing with the problem posed by inaccessibility.

One of the most significant of these corollary questions is: Are the dimensional properties of relativistic cosmology (e.g., homogeneity and isotropy) that I have raised here actually relevant to our analysis of psychology? A part of the answer to this question, as I have already hinted, is that expansions of the Newtonian world such as relativity and quantum theories are extensions that add to and build on the world of our scale. Thus, at the very least, there can no more be allowable violations of the laws of cosmological physics by mental processes than of our local Newtonian-scale world. Any situations in which mental processes conflict with physical laws and properties at any scale and of any kind, therefore, must be considered to be further support for the argument that they may be outside the domain of ordinary science.

In short, when there is a conflict between the most basic properties of mental and physical processes, it is not just a matter of differences in content or immaturity of the science. Rather, the entire complex of ideas that allows us to use the logical, inferential, axiomatic-deductive system that characterizes physics may be inoperative for psychology. We have to assume that whatever the distorting forces are that account for these differences, they are not understandable using the methods that have evolved to meet the needs of physical science. What is inaccessible remains inaccessible because there is no psychological equivalent of the Cosmological Principle that "laws there are the same as here" for psychology! Nevertheless, they are a part of the physical world, albeit a part that requires a different scientific approach.

Finally, I have to reiterate a comment that I have had to make many times over the past decades. In spite of the fact that it may not be possible to define the mind or to access mental processes and mechanisms, it is necessary to use a mentalist vocabulary throughout any discussion of this kind. This book, like all other efforts in communication, oral or written, is aimed at transferring viewpoints, ideas, and concepts between "minds." It is a presumption of any such communication that minds exist, however difficult they may be to define. There is an unavoidable necessity, therefore, to use words that are loaded with connotations and ill-defined denotations. Without such tools we could not communicate with each other at all. I find it necessary, therefore, to include mental terms in my vocabulary. I cannot avoid referring to hypothetical constructs such as mind, cognition, cognitive processes and mechanisms simply because there is no other way to do it, other than rejecting the reality of mind—something I have not done anywhere in my writing. What we cannot do, I argue, is to reduce these mental terms to specific entities, cognitive modules, or neural mechanisms and to measure them and draw inferences about these hypothetical constructs in the same way physicists deal with their more tangible objects and events. That is the curse of inaccessibility and the fundamentally ill-posed, underdetermined nature of the answers we get in psychological research. Indeed, I must assume that there are some hidden and inaccessible processes going on within the brain to explain why something different is coming out than went in. These cognitive processes and mechanisms are what are inaccessible, not the stimuli or the behavioral responses. This is the way psychology has always been, and it is likely that this is the way it will always be. The goal of this book is to explain why this is the case.

The remaining chapters of this book are organized as follows. Chapter 2 deals with one of the most important areas of psychology research: the quantifiability of its dimensions and properties. In that chapter I show that there seems to be a disconnect between the basic arithmetic properties and the way that humans deal with numbers.

Chapter 3 reviews the behavioral literature to demonstrate that in many cases humans behave quite differently than would be predicted by the properties of physical time and space.

Chapter 4 examines the reasons that psychology dotes on statistics, whereas physics is best served by conventional analytic mathematics. I attribute the primary difference to the fact that the properties of physical and psychological dimensions differ. A secondary reason is that analytic mathematics and statistics differ at best in what they can accomplish. In particular, the inability of statistics to root out causes (as opposed to describing correlations) also contributes to the fact that most psychological theories are statistical. The thesis of this chapter is that statistics and psychology fit well with each other, and analysis and physics fit well, but that psychology and conventional math are disconnected by their respective properties.

My goal in Chapter 5 is to summarize the discussion presented in this book that shows that the properties of psychological activities account for the inaccessibility of mental processes and mechanisms. I argue that this inaccessibility is fundamental and prevents us from answering some of the grand questions that have traditionally been asked by psychologists. It is in this final chapter that I also suggest the shape of a useful scientific approach to psychology. Its name is familiar—behaviorism. My approach is, however, perhaps even more radical than some of its predecessors.

NOTES

[1]Cognitive penetration is defined as the distortion of our thoughts and percepts by logical or emotional influences beyond our control.

[2]This idea of multiple universes is an extrapolation from string theory discussed on page 11. It is referred to as the "landscape" version of string theory. A major problem with landscape theory in describing the properties of multiple universes is that its equations are likely to be unsolvable; that is, it may represent an NP complete problem. (See Minkel, 2006, for a compete discussion of this problem.) A further problem is that the intellectual foundation of multiple universes—string theory—itself has come under serious criticism recently (Smolin, 2006; Woit, 2006).

[3]The implication that the red shift is not determined by the relative velocities of distant objects but by some kind of photonic fatigue would be enormous. Virtually all of our cosmology is based on this phenomenon and would have to be restructured if it was proven to be incorrect.

[4]Even in this brief paragraph there are mentalist terms that could lead a careful reader to fairly ask such probing questions as: "What do you mean by that word?" Terms such as response, behavior, cognitive, mental, and so on, are still the objects of contentious debate. I do not wish to spend any more time futilely trying to provide readers with precise definitions of these elusive specters. Let's just use their commonsense meanings so that we can carry on this discussion. Otherwise, we would not be able to communicate at all.

[5]Webster's International Dictionary has a column and a half dedicated to definitions of time; all are equally unsatisfactory.

[6]We can begin to discern trouble for our effort to appreciate the meaning of psychological time in this definition with its allusions to "continuity" and "irreversibility." As we see later, psychological time often appears to be discontinuous and, paradoxically, often to be reversible.

[7]Newton also alludes to "relative" time in this same quotation, but it is not in the same sense as the relativistic physical time of the twentieth century. His comment was:

> relative, apparent, and common time, is some sensible and external (whether accurate or unequable) measure of duration by the means of motion, which is commonly used instead of true time; such as an hour, a day, a month, a year.

His allusion to "apparent" and "sensible" suggest he was referring to psychological time as distinct from "true [physical] time."

[8]The Lorentz transformations for both time and space were originally formulated by Joseph Larmor (1857–1942) in 1897, two years before Hendrik Lorentz (1853-1928) also published them based on electrodynamic considerations. It was Einstein's great achievement to use his Special Theory of Relativity to derive these laws from basic principles of space and time.

[9]The tri-dimensionality of space is no longer taken for granted by string theorists. The "threeness" of space seems to be related to the intuitive structure humans have of the world in which they live. However, as already noted, multidimensional string theory has come into question recently.

[10]Einstein received the main credit for the Special Theory of Relativity that was developed at the beginning of the twentieth century. However, there was considerable ferment even then concerning the priority of the relativity ideas. Others, such as Poincaré, were articulating nearly equivalent versions. According to Isaacson (2007), Poincaré never really understood Einstein's formulation and involved superfluous postulates in what was a simpler formulation. Einstein, however, was the one whose formal model best described and integrated the many different ideas involved in this momentous development. Although there is still considerable debate about priority, Einstein seems to have brought it all together in what is now accepted as its current form.

[11]Of course, it is possible to have scales that are not equal-interval or linear scales. For example, logarithmic scales are frequently used to compress data, so that small as well as large values can be represented on the same graph. However, this is only a convenient way to plot data and does not alter the basic properties of time and space that are being measured.

[12]In other instances, Einstein argued that space-time was continuous since the points x, y, z, t could be arbitrarily close together (see p. 51 of Einstein, 1917/2003).

[13]It should not be overlooked that a kind of time travel is possible in accord with the dilation of time (see Equation 1.1) at relativistic velocities. Should we be able to travel at speeds approaching the speed of light, time is altered with regard to other observers in a way that would allow travelers to return to their point of spatial origin at a later time for the origin than for the traveler. The high-speed traveler might have passed only a few years, while those at the point of origin experienced much longer periods—the exact amount depending on the velocity of the traveler. However, there is no equivalent explanation of how we might reverse time and arrive at a time earlier than when one departed.

[14]In recent years, the issue of time invariance has become somewhat more controversial with experimental observations—the decay of a particular variety of meson—suggesting that this concept is not correct. Some authors believe that this raises doubts about the invariance of physical events with time reversal. Sachs (1987) deals with this issue in detail.

[15]However, there may be local decreases in entropy due to energy-consuming processes, most notably living things. This does not challenge the basic idea that, for the whole universe, entropy must increase.

[16]With thanks to my colleague Peter Killeen for calling Aristotle's four causes to my attention.

[17]In Chapter 4, I discuss the special role of statistics, a subfield of mathematics. Here, I am concerned with the role of conventional, analytic mathematics, also to be defined later.

[18]Not all of it, of course. Some pure mathematics was developed before the application to which it was eventually attached. Even then, the newer math was based on older premises that had been, in turn, based on physical relationships.

[19]Not all of mathematics is driven by physics. Some mathematical concepts deal with arcane matters such as the meaning and nature of number systems that may have initially seemed to have no physical relevance. However, much of mathematics has evolved to serve the needs of physics.

[20]This last comment is especially important; it suggests that conventional mathematics and statistics are incompatible, not only so far, but also into the foreseeable future. Since psychologi-

cal data are wallowing in "error" (i.e., variability), Luce hints here at a barrier between psychology and conventional mathematics and, therefore, between psychology and physics comparable to the thesis of this present work.

[21]I am aware that physicists also use statistical techniques and methods to deal with especially complex problems that in some cases are comparable to those challenging psychologists. Many of these complex problems are as intractable to conventional analysis as are the psychological ones. It should also be pointed out that even the statistical approach requires the use of the mathematical primitives now to be described. Later I discuss how statistics may have been developed as an expedient to overcome the inapplicability of conventional mathematics, and how fundamentally different the statistical approach is from the conventional mathematical one, as well as how much more appropriate it is for modeling psychological systems.

[22]Basic arithmetic is sometimes expanded to include other aspects of number theory beyond these four simple operations.

[23]It is appropriate to point out here that many useful measurements are made with scales that are not meaningful. For example, we use the Fahrenheit and Centigrade scales to measure temperature all the time. However, these are interval scales that have arbitrary zero values. This means that the values of degrees C and F, respectively, are not meaningful in the sense suggested by Suppes and Zinnes (1963) and Falmagne (2004). The use of these scales is a convenience to keep our numbers small and to link them with the freezing point of water, but they do not have the same mathematical power as the Kelvin scale, which does have a non-arbitrary zero. A degree on the Centigrade or Fahrenheit scale does not tell the investigator how far above the true zero the temperature is; a degree on the Kelvin scale does! Even more important, the Kelvin scale is rife with clues to the meaning of temperature, since 0 deg K means that all motion has ceased.

[24]Coombs (1950) went on to describe how psychological scaling could be done without a unit of measurement. He was interested in emphasizing the ordinal effects in ranking experiments. I am concerned here only with the traditional definition involving quantitative effects.

[25]The Cosmological Principle has received empirical support from the even distribution of cosmic background radiation. The accidental observation in 1965 by Arno Penzias and Robert Wilson remains the main empirical argument for what had hitherto been a philosophical extrapolation of Copernicus' assertion that the earth was not the center of the solar system.

2

Cardinality, Measurability, and Quantifiability of Psychological Phenomena

2.1 INTRODUCTION

This chapter is concerned with the most basic attribute or property of psychological functioning—its quantifiability; that is, its ability to deal with and to be described by numbers. In other words, I now ask: Do the basic numerical and arithmetic properties that were identified as requirements for robust measurement and quantifiability for the physical dimensions in Chapter 1 apply to psychological dimensions?

Before beginning, I should note that there is considerable popular opinion that psychology is inherently unquantifiable. Some, mostly lay persons unacquainted with the high level of methodology developed for psychological research, assert that since it is a "human science," the entire effort to imitate physics and measure mental processes in the laboratory is a wasted one. I want to emphasize that this is *not* the argument I am making here. Recall the purpose of this book: It is to consider whether or not the accessibility barrier can be overcome for psychology in the same way it was for physics. In other words, does the domain of psychology exhibit the same properties of space, time, and quantifiability that permits it to mimic the inferential and deductive successes of physics? This is not the same thing as denying the value of psychological experimentation. As a matter of fact, most behavioral observations made during an experiment are unquestionably quantifiable. It is what we do with them—our "just so" inferences from behavior to hypothetical constructs—that is the topic of this book.

The specific problem I confront here is: Do the differences between the two sciences—mainly the lack of any general principle that the laws and dimensions of cognitive processing are insufficiently uniform—make the difficulties faced by psychologists who assert that mind can be *inferred* from behavior insurmountable? In other words, is there some kind of a psychological equivalent to the Cosmological Principle[1] that permits us to deduce mental processes (there) from behavioral data (here) in the same way that physicists can infer the properties of distant objects from observations of the signals they send?

Whatever the outcome of this debate, it must be kept in mind that our choice is not between "No Psychological Science" and "Science of the Usual Kind." Instead, it is between different kinds of science, one based on the possibility of reductive inference (cognitive mentalism) and the other denying this strategy (behaviorism). The history of psychology is replete with examples of the stress that this basic issue imposes on our science.

There are many kinds of possible psychological scientific strategies. Perhaps the most consistent and logical one ultimately will have the operational and descriptive properties of a classic behaviorism; perhaps not. Perhaps it will have statistical properties of stochastic models; perhaps not. In no way, however, am I even suggesting that all of psychology is beyond the pale of scientific inquiry. At the very least, measurements of overt behavior are strictly quantifiable, and that includes almost all laboratory findings.

The main thesis of this book, however, is that the mechanisms and processes of the mind cannot be inferred from observable behavior in the same way that the properties of distant galaxies can be inferred from what are largely indirect spectroscopic measurements. This thesis is based on the following arguments:

- The one (behavioral observation) to many (possible mental mechanisms) rule.[2]
- The cognitive processes underlying behavior are inaccessible to standard laboratory techniques of measurement and quantification.
- The properties of psychological time and space are not the same as those of the physical world.
- We do not have the advantage, therefore, of saying that the laws of the mind are the same as those of overt behavior.
- Therefore, it is not possible to infer from behavior what are the underlying mental processes and mechanisms.
- Therefore, the laws of mental causality are extremely difficult if not impossible to identify. Only descriptive correlations are possible.

- Unfortunately, whatever correlations may be observed between environmental causes (stimuli) and behavior do not allow us to impute or infer unique mental causes (Yule, 1926; Cheng, 1997).

Another caveat that has to be expressed before I continue this discussion is that by no means am I rejecting the reality of mental processes. Although it is difficult to define the mind, and standard measures of mental activity are always open to judgments of validity, there seems to be little argument that our consciousness or self-awareness is a real and natural result of neural activity. No one yet knows how the mind is generated by the brain, but there is hardly any disagreement that mental activity is a process of this particular material entity. An analogy that I have used before is that "mind is to the brain as rotation is to the wheel." To reject the reality of mental processes would make all human activities meaningless, just as to deny rotation would make the concept of circular motion meaningless.[3]

In sum, the essential problem of measuring mental entities is beset with difficulties, including the inaccessibility of these mental processes, the absence of a bridging principle comparable to the Cosmological Principle, the effect of that absence on our ability to infer mental structure, and the lack of conformity with certain criteria for quantifiability. The remainder of this chapter discusses a group of reasons for accepting the questionable nature of human measurement.

The argument presented here is not, I should note, a new criticism of psychology. In a wonderfully insightful treatise on measurement, Hand (2004) discussed the long history of controversy that surrounds measurement in psychology. He directed our attention to a number of critical comments about psychological measurement, one of which is especially salient to the argument being made here. Dawes and Smith (1985) are quoted as follows:

> It is not uncommon for psychologists and other social scientists to investigate a phenomenon at great length without knowing what they are talking about. So it is with *attitude*. While 20,209 articles and books are listed under the rubric 'attitude' in the *Psychological Abstracts* from 1970 to 1979, there is little agreement about the definition of *attitude* and hence what aspects of attitude are worth measuring. (p. 509)

This admonition certainly holds true for all other mental processes, including the most superficially simple psychophysical functions as well as the far less concrete measures found in fields known as psychological measurement of intelligence, personality, and abnormal mental states.

2.2 ON CARDINALITY

The best place to begin this discussion of the quantifiability of cognitive processes is at the beginning; that is, at the most primitive kinds of quantitative functions and processes. Nothing fits this criterion better than the way in which cardinal numbers are assigned to events and objects by human beings. Cardinality refers to the conceptual equality of a numeral and the number of objects in a set. It is the most primitive form of what Stevens (1951, p. 23) referred to as "the isomorphism of numerals and objects or events." This isomorphism can come in many forms, but all of them seek to assign numerals to phenomena. In fact, the assignment of numbers to phenomena was all that Stevens asked of measurement. This misinterpretation of the meaning of measurement continues to plague psychology, as has been pointed out by Michell (1999) so effectively.

In the following discussion, my goal is to show that human beings do not automatically associate numbers with numerousness in the same way that cardinal numbers are supposed to work. There is little question that cardinality is implicit in the real world and in the physicists' interpretation of it; a hundred stars has the same cardinal meaning as a hundred golf balls. The major conclusion to be drawn from this brief review in this chapter is that, at the human cognitive level, there are major discrepancies between our perception and appreciation of the number of things and the actual number of things.

To examine the limits of human cardinality, it is useful to start by exploring the abilities of children and animals that have not fully developed the higher concepts of numbers and their applications. Children seem to be able to serially count a set of objects before they understand the significance of the last number in their counting: the cardinal number that is equivalent to the number of objects in the set. That is, it is more difficult for youngsters to understand the significance of this last count as a representative of the number of objects than it is to count the number of objects. Mix, Huttenlocher, and Levine (1996), for example, showed that children acquire a preexisting counting ability before they can compare the numerousness of items from two different modalities—a sequence of drum beats and a collection of visually presented objects. Since counting skill emerges only after 3 or 4 years of age (Piaget, 1952; Wynn, 1990), it likely that cardinality must also be delayed to at least this age. Clearly, cardinality is missing from the youngster's repertoire of mathematical skills until a certain level of development. Any assumption that a cognitive awareness of the concept as well as the appreciation of cardinality underlies their behavior before that age would appear to be incorrect.

Surprisingly, although slightly older children may not yet have developed a sense of cardinality, a primitive sense of ordinality may occur even earlier at 2 or 3 years. Brannon and Van de Walle (2001) showed that children this young can respond in a way that indicates that they can tell that some groups of objects are larger or smaller than some others—if the numbers are relatively small. However, these researchers suggest that this is not because of some numerical competence, but because of an analog ability to estimate "continuous" amounts independent of the actual discrete number of objects. In other words, these critics are suggesting that it is not the numerousness of the objects but the physical size of the cluster of items that is the distinguishing cue in this early appreciation of ordinality.

Brannon and Terrace (2000) have further shown that rhesus monkeys can order groups of objects at least up to the number 9. That is, they can tell that one group of objects contains more objects than a smaller group if the number in the larger group is less than about 9. This can be done for several groups of objects if presented in order. However, when a nonmonotonic (disorderly) sequence of groups of objects was used, the monkeys were not able to order the group sizes. This result suggested that they were limited in their use of cardinal numbers, even though they did appreciate the concept of ordinality.

Thus, Rhesus monkeys, like young children, are apparently also able selectively to respond in some primitive fashion to ordinality expressed as an ability to compare stimuli that are "more than" or "less than" each other (Hauser, Carey, and Hauser, 2000). This ability seems to function only if the comparisons are limited to relatively small numbers.

There is, therefore, an important alternative explanation of these findings. The specific behavior (e.g., distinguishing between two groups that physically differ in the respective number of objects) can be accomplished by two quite different cognitive strategies: counting and size estimation. One of these processes (counting) is essentially a discrete process involving the apprehension of numerousness; the other (size estimation) is essentially an analog process involving the perception of group size differences. This uncertainty is an example of the problems caused by the one (phenomenon) to many (explanations) constraint discussed earlier. It is unlikely that this controversy can be resolved in the future, and even more likely that some third (or fourth or fifth) alternative might be suggested. This kind of irresolvable debate is typical of many psychological controversies. It is also a straightforward example of how difficult, if not impossible, it is to infer underlying cognitive mechanisms from observed behavior.[4]

We may conclude from these studies that different quantitative skills and concepts emerge at different points in the developmental and evolutionary cycles. The question now arising is: Just how proficient do we become dealing with numerousness, (especially cardinality) as we mature? The answer is surprising. It turns out that our ability to deal directly with cardinal numbers remains quite limited as humans develop. Although it is possible for adults to exhaustively count, however ponderously and however prone to error, to very large numbers, the intuitive appreciation of specific numbers falls off at an astonishing small number—just a few more or less than 7—depending on the argument so influentially expressed by the distinguished contemporary psychologist George A. Miller (1956).

2.3 SUBITIZING

Subitizing is the "direct apprehension" of the number of objects without explicit counting. It was probably first described by Jevons (1871), but the use of modern term *subitizing* is usually attributed to Kaufmann, Lord, Reese, and Volkman (1949).

Subitizing is limited to a relative small number of objects. A substantial body of research has suggested that we can "directly apprehend" the numerousness of groups of objects when the numbers involved are less than 7. Beyond 7, the ability severely declines, and reaction time to answer a question of how many briefly exposed objects were present drastically increases just as the accuracy decreases. Beyond 7, the apprehension of numerousness requires that we indirectly estimate or even count the objects to get an apprehension of quantitative value. The apprehension or attachment of cardinal numbers to sets of objects beyond this relatively small number, therefore, is not a "natural" property of human mentation. This is one of the first glimmerings that there is not a good fit between the mathematical and psychological concepts of cardinality.

The interpretation of the subitizing phenomenon is that the innate quantitative abilities of most people (with the exception of a mysterious few[5]) are very limited. Not only do they develop fairly late in our childhood, but at best we do not exhibit the tight relationship between cardinal numbers and the number of objects in a group that is characteristic of the physical sciences even as adults.

Furthermore, the larger the number, the more difficult it is to discriminate differences. It is easy to determine that sets containing 5 and 6 items are different. However, it is impossible for us to determine that sets containing 1,000 and 1,001 objects are different. Our personal number systems collapse at such large numbers; indeed, as we now see, they do so at much smaller numbers. One implication of these findings is that the psychological distance between large numbers and small numbers is not the same; there is no property comparable to physical homogeneity present in psychological space. Psychological numerousness is, therefore, neither equal interval nor homogeneous. Even worse, as we discuss in the next chapter, the ordinality so characteristic of number systems does not seem to be maintained in human cognitive processes.

Quantitative physical dimensions, to the contrary, are precise to limits defined by the effort used to measure them down to the Heisenberg limit; there is a very tight natural relationship between numerousness, cardinality, and the other properties of mathematics and physical dimensions. Whatever this relationship is (and set theorists still debate esoteric mathematical theorems concerning it), there is no question that universal cardinality is a basic foundation of the mathematical system that is rare or nonexistent in human mentation.

The most extreme insensitivity to numerousness by humans can be found in studies of preliterate people. Two groups of Amazonian tribal people have been studied in recent years. Gordon (2004) spent time with the Piraha people of the Amazon basin and discovered that they had only a few words for numbers including 1, 2, and many. In simple tests (e.g., match the number of items in two rows) matching performance was very poor for numbers greater than 3. Thus, these people seemed to have very little sense of numbers either in the sense of ordinality or of cardinality.

Another Amazonian tribe, the Munduruku, studied by Pica, Lemer, Izard, and Dehaene (2004), had no words for numbers greater than 5. These people seemed not to be able to subitize or immediately apprehend numbers beyond 3 or 4, a value just half of that reported for their counterparts in America and Europe.

Although both research groups believe that the people who lived in these simple societies could not count beyond small numbers, they did seem to have some kind of approximate (analog) appreciation that some groups of things were greater than others. Furthermore, even though both of these tribes exhibited very poor counting and arithmetic abilities, the Munduruku were very competent in appreciating geometrical concepts. Dehaen, Izard, Pica, and Spelke (2006) presented simple maps using geometrical properties to the Munduruku people. The results suggested they had well developed concepts of lines, parallelism, distance, and angle available to them. In retrospect, this may not be too surprising considering that these people, like all of the rest of us, live in a geometrical world, and none of these concepts would have been alien to them. On the other hand, as far as arithmetic goes, the reduced needs for counting in these Amazonian cultures seem to have inhibited not only their use of number words, but also their ability to deal with some of the basic properties of numbers and arithmetic.

These interesting results suggest that at least some humans do not have the most basic cognitive skills necessary for the quantification of their world. How general this failure to deal directly with quantity may be for the rest of us is discussed in the next section, the summary point being that there is a huge divide between physical and psychological numerousness or cardinality, one of the basic properties on which measurement must be based.

2.4 THE MAGICAL NUMBER 7 (PLUS OR MINUS 2)

Are the limited numerical skills of children, monkeys, and Amazonian people just quaint curiosities? Or, to the contrary, do even citizens of our most advanced cultures exhibit similar constraints on their appreciation of cardinal numerousness? Just how limited are the numerical powers of human beings was made famous by the ever-popular article, "The Magical Number Seven, Plus or Minus Two: Some Limits on our Capacity for Processing Information" (Miller, 1956). Miller's goal was to evaluate the limits of human direct numerical processing. Although he did not specifically mention cardinality in his article, it clearly was the basic issue under examination. Indeed, direct apprehension, cardinality, and absolute identification all probably represent closely related, if not identical, aspects of the same fundamental process.

Miller built his ideas on the formal development of information theory as it was defined in the newly emerging communication field. He cited a number of what are now classic psychological studies of information transmission by humans, all of which showed a marked deterioration in absolute numerical judgments above a relatively small number: "seven plus or minus two." The experiments he summarized also showed that this constraint occurred only when information was presented along a single dimension; it could be overcome and more information transmitted by providing additional dimensions or by simultaneously adding a second or reference value to the displayed stimulus. Rather than recapitulating his discussion, I now just summarize in the following list a few of the experiments he cited:[6]

- Pollack (1952) showed that most people could not absolutely identify more than six different tones by name with a high degree of success. (However, see the subsequent discussion on absolute pitch for a modern elaboration of this finding.)
- Pollack (1953) also showed that the limit of six tones could be exceeded if we provided additional dimensions. For example, tones that varied in both frequency and intensity permitted more than six tones to be absolutely identified. Indeed, if some kind of pattern was provided to describe the interactions of the two dimensions, then more than eight different combinations could be absolutely identified. It is clear that providing some structure to the test stimulus increased the absolute identification (naming) capability of the observer.
- Klemmer and Frick (1953) also showed that the limit of approximately six or seven items could be increased if we provided subjects with more than a single dimension to evaluate. For example, people could absolutely identify which of 24 positions was marked in a square array given the presence of two well-ordered spatial dimensions. A square, of course, introduces a kind of regularity that, in effect, restructures or recodes the problem into something that may actually reduce the task to the single-dimensional one. Miller also suggested that our ability to recognize large numbers of things like faces may be due to the addition of multiple dimensions.
- Garner (1953) showed a similar breakdown in our ability to absolutely distinguish between different sound levels for more than five intensities.
- Miller cited work by Hake and Garner (1951), as well as some unpublished data by Coonan and Klemmer, who explored our absolute perception of visual position that suggested that there are "between 10 and 15 distinct [discriminable] positions along a linear interval." Miller noted that this is the largest number of absolute identifications ever "measured for any unidimensional variable."

The general conclusion to which Miller came was that people have a very limited ability to deal *directly* with numbers; that is, to directly perceive numerousness. Whether measured in bits (approximately 2.5) or integers (approximately 7), it was clear that our ability to process number is severely limited unless we are provided with additional information or dimensions. The general result obtained by all of the experiments mentioned in Miller's famous article is summarized in Figure 2.1. As Miller (1956) suggested, there is a leveling off at about seven input

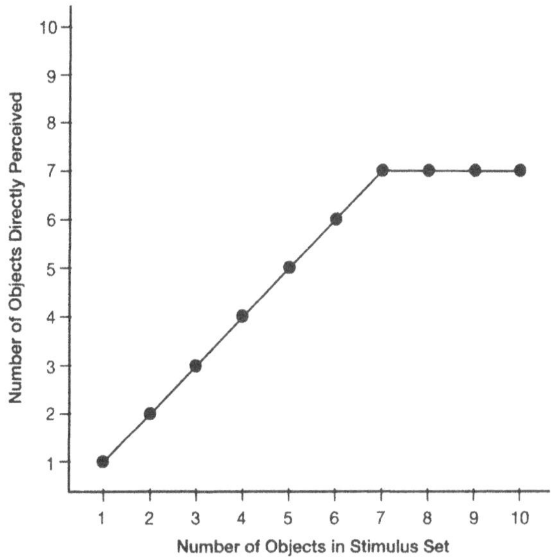

Fig. 2.1 The limits on direct numerical apperception.
On the average, subjects are not able to directly perceive the number of items in groups larger than seven (plus or minus 2).

stimuli in our ability to be immediately aware of the number of stimuli when we are limited to a single dimension.

The important point made by Miller's review (beyond the bare empirical facts) is our limited ability to directly deal with numerousness. That is, we do not deal with numbers in an isomorphic fashion in which our perceived perception of numerousness directly corresponds to quantity. There is, on this interpretation, a considerable lack of conformity between quantity in the external physical world and the limited way we perceive numerousness as measured in these psychophysical experiments. This limited human ability to deal with numerousness demonstrates the failure of one of the most primitive forms of number properties: cardinality. It is another example of how human mentation differs in its most fundamental principles from some equally fundamental principles of mathematics. This difference is additional evidence that there is no psychological equivalent of the Cosmological Principle.

2.5 COLOR AND PITCH DISCRIMINATION[7]

Our ability to count and to distinguish differences in numerousness, as we have just seen, is severely limited. Furthermore, we are insensitive to other well-defined properties and metrics of a variety of other physical stimuli. A further case, therefore, can be made for the idea that the laws governing the quantitative processes going on inside our heads may be quite different from the mathematical processes of the external physical world. To a surprising degree, psychological properties are not congruent or isomorphic with the dimensions, metrics, or the properties of the physical world. These nonveridicalities (disagreements between the stimuli and reports of what we perceive) are among the reasons that inference from behavior to mental phenomena is fundamentally blocked, and that the mind remains inaccessible to direct experimental assay.

This barrier to inference does not necessarily mean, as some would argue, that our behavior need be maladaptive because of these inconsistencies. We do respond appropriately to most of the challenges of the world. However, on close examination, the many nonveridicalities between stimuli and responses suggest that we do some major recoding of the stimuli in a way that makes the properties of the internal processes quite different from those of the externally observable behavior. This cryptic recoding is the topic of the next section.

Let's start this discussion with one of the most basic observations of human information processing. Often overlooked in any discussion of mental "units" is the basic fact they are not equal-interval scales. Rather, the size of the stimulus producing a "unit" of psychological experience often changes with the intensity of the stimulus. Furthermore, even in those cases in which equal intervals may appear, this does not guarantee that equal ratios also be characteristic of the mental dimension. The quantifiability (in Coombs' sense; see page 26) of the underlying cognitive dimension, therefore, becomes questionable.

The point being made is that one of the most basic ideas that made inference possible from observation to structure in the physical world—homogeneity—does not hold for psychological processes. Nowhere is this more evident than in the variation in the size of the just noticeable difference across many different dimensions of human sensory experience.

The general law that testifies to this failure of our sensory systems to behave homogenously is what was perhaps the first psychophysical relationship to be expressed mathematically: Weber's Law[8]:

$$\frac{\Delta I}{I} = C \qquad \text{Equation 2.1}$$

where ΔI is a just noticeable difference in a stimulus, I is the intensity level of the stimulus, and C is a constant. Obviously, this expression means that the physical stimulus unit size associated with subjectively constant intervals increases as a function of the stimulus intensity.

Although it may be argued that this kind of compression takes place in physical as well as in cognitive processing, there is a major difference. Plotting some function on a nonhomogeneous scale (e.g., a logarithmic axis) can be a convenience for physicists; it permits them to deal with large numbers as well as small numbers on the same graph. For psychologists, however, it is much more fundamental; the actual dimensions of the mind seem to vary in irregular ways. Weber's Law, for example, is a description of the psychological properties of the human. No convenience here. Rather it represents a mental violation of one of the main criteria of quantifiability: equal-interval scales.

On the other hand, no one argues that the units of time and space are not constant in distant parts of space, subject to relativistic effects, of course. However, the units of human time and space seem to be flexible depending on their magnitude. Weber's Law, therefore, is among the most general expressions of the nonhomogeneity of a psychological dimension and the nonveridicality of physical and psychological dimensions.

Another pair of examples of how flexible are human discriminative processes can be found in the phenomena of color and tone or pitch discrimination.[9] Both points in the CIE color space (defined by the wavelength of the stimulating light) and acoustic experiences (defined by the frequency of the oscillating pressure waves) are perceived as continuous functions. That is, we do not have any awareness of any discontinuities when we examine a spectrum of chromaticity or tonality. Nevertheless, if asked to discriminate between two different colors or two different tones, most people report an extremely fine ability to tell that two adjacent stimuli are different. For certain photic wavelengths (e.g., 500 and 600 nm) the differential threshold for visual stimuli may be less than 1 nm (Judd, 1932). Since there are other dimensions of color perception (including contrast, saturation, and intensity) it is likely that we can *discriminate* tens of thousands, if not millions, of different pairs of colors from each other when the pairs are presented simultaneously for direct comparison.[10] Similarly, people can discriminate very well between two nearby acoustic frequencies when the two are presented close in time to each other. However, when a single color patch or acoustic frequency is presented, we do very poorly at naming it. Indeed, we use only very few name categories to describe wide ranges of possible chromatic and tonal experiences. This ability is referred to as *absolute identification*. Let's consider vision first.

For many years, it has been generally agreed that only a small number (10-12 and perhaps exactly 11) of color names are used to represent the many colors that compose our phenomenological color space when absolute judgments are required. Furthermore, it was proposed that this small number of categories was the same from one language, culture, or group to another (Berlin and Kay, 1969). The color names, although differing in different languages, did seem to be associated with a few privileged points in the color space. In English we call these 11 privileged points red, yellow, green, blue, purple, brown, orange, pink, black, white, and gray (Kay and Regier, 2003). Many languages use different names but, whatever the language, comparable words seem to be associated with these privileged points.

In sum, we have only a modest ability to absolutely identify colors, just as we have a modest ability to absolutely identify numerousness. Furthermore, the color names are not evenly spaced on the wavelength dimension. Yellow is the name given to color experiences produced by wavelengths that vary from about 570 to 580 nm; orange is the name given to wavelengths that vary from about 580 to 610 nm; and green is the color name given to the experiences produced by wavelengths varying from about 500 to 550 nm. The point is that the units of absolutely discriminable color experiences are also not isomorphic with the units of the physical stimulus producing them.

There is still considerable controversy about the origin and nature of this privileged set of color names, especially concerning their exact number and degree of universality. Of particular interest has been their possible relation to the underlying photochemical and neural mechanisms that might possibly underlie this constraint on our color naming abilities. No absolute identifications between color names and retinal photochemicals, however, have ever been successfully made. Despite this theoretical lacuna, the basic observation (the behavioral fact) is valid; only a modest number of categories are used by people when confronted with the task of absolute color naming. Thus, in the absence of a comparison standard permitting fine discrimination, the phenomenological units of color perception change size, just as they do in the intensity domain. Therefore, there appears to be a breakdown in color perception in what physicists have referred to as homogeneity—the evenness of the intervals—along the color experience dimension. There may be no good physiological explanation of why this should be the case, but there is no question that the homogeneity of intervals that helped physicists make sense of the cosmos does not hold in the subjective visual domain.

For acoustic stimuli, frequencies differing by as little as 2 Hz may be distinguished, (i.e., reported as conveying different pitches; Shower and Biddulph, 1931) if one is presented immediately after the other. Furthermore, as in the analogous situation for vision, if we provide additional information such as simultaneous variations in their intensity or overtones, thousands of different tonal pairs can be discriminated from each other. However, as in vision, if a single tone is presented, people do very poorly at absolutely identifying it.

Although most people are not capable of assigning a specific name or even a general category to a tone, there are a few blessed with what has been called *absolute pitch*. The best of these gifted individuals can assign the correct musical name to an isolated frequency for as many as 70 or 75 musical tones (Ward, 1963; Levitan and Rogers, 2005), but they are rare. Estimates are that only 1 in 10,000 people exhibit anything close to even this limited definition of absolute pitch (Ward, 1999). Even then, it has to be appreciated that absolute pitch is a relative term and does not mean that the absolute naming capacity comes anywhere close to the number of pairs of tones that can be discriminated from each other.

As Levitan and Rogers (2005) and Deutsch (2006, personal communication) have emphasized, absolute pitch comes in many guises; some people are able to identify only a single tone, and others can identify several but make a variety of different kinds of errors confusing, for example, tones from different octaves.

Furthermore, absolute pitch seems to be very susceptible to training and life experiences. Deutsch, Henthorn, and Dolson (2004) showed that speakers of languages that depend on the tonal quality of the enunciations (e.g., Vietnamese and Mandarin) have heightened absolute pitch capabilities that approximate what some might call absolute pitch. Deutsch, Henthorn, Marvin, and Xu (2006), among others, have proposed that the *potential*, if not the *actualization*, for absolute pitch is, therefore, much more widespread than currently believed. In fact, they suggest that the propensity may actually be universal and can be developed if children are exposed early to situations in which they must assign names to tones. Because so few people (Americans in particular) have this opportunity, this ability never develops and is lost in most adults.

Once again, the major point is that our tonal experiences, like our visual ones, do not appear to enjoy the same kind of homogeneity or equal-interval properties on which are based all physical theories. It is interesting to note that this same failure to discriminate tone or hues is also reflected in something more global: face discrimination and recognition.

2.6 RATING AND RANKING[11]

Another instance in which our discriminative abilities are severely limited can be found in situations in which we are called on to place items in a group in order along some dimension. There are a number of methods by means of which items can be placed in order. If simple normative or verbal tags are used, the procedure is referred to as *rating*; if ordered numerical values are applied, the process is referred to as *rank ordering*. In point of fact, the difference is inconsequential. Rank ordering is almost as completely non-quantitative as is simple verbal naming. Both methods unjustifiably suggest a level of quantifiability since the numbers used in rank ordering are really just "names" attached to an ordered sequence, as are verbal phrases such as "more than," "less than," or "equal to."

The technique of rank ordering a set of stimuli is one of the most common techniques in test evaluation. Unfortunately, the resulting ordinal scales of measurement exhibit few of the properties required for true quantification. Although, there are perfectly useful statistics (e.g., the Spearman rank: order correlation coefficient) for comparing two ranked series of objects or events, there may be serious discrepancies among the intervals between items in each set. Two different rank orderings may completely obscure the nature of the actual measurements (for example, the range of a measured variable) and imply quantifiability when, in fact, none of the properties required for true measurement is present (see page 28).

To clarify this problem, consider that five items ranked in order from one to five, but which vary in magnitude on some physical scale from 20 to 30 on some interval or ratio scale, may (rank order) correlate perfectly with another set of five items placed in rank order that vary in magnitude from 50 to 100. In this manner, similarity may be falsely attributed to the two series. The rank order correlation, in other words, implies a relationship where none may actually exist.

Such a misunderstanding may have profound ramifications not just on scale and order judgments but, in some situations, can actually reverse one's general conclusions about an experiment. Bartoshuk and her colleagues (Bartoshuk, Fast,

Duffy, Prutkin, Snyder and Green, 2000; Fast, Green, Snyder, and Bartoshuk, 2001; and Carpenter, 2000) have recently brought to our attention a very serious problem that can occur with careless use of this kind of ranking scales. The work of Bartoshuk's group is centered on taste perception; one of the main goals of research in this field of sensory science is to quantify the relative "strength" of the gustatory experience produced when chemicals of different concentrations are applied to the tongue. One tradition in this field has been to simply ask observers to rank order their experiences by using words such as "weak," "moderate," and "strong."

Bartoshuk and her colleagues demonstrated how insidious the effects of such free-floating standards can be. They pointed out that when the different rating scales of a group of observers are pooled, it can actually lead to what they refer to as a "reversal artifact" that can completely reverse the conclusions drawn from even the best designed experiment. Furthermore, a conclusion of "no effect" can result from the erroneous assumption that such a judged scale value as "very strong" means the same thing to all observers.

Bartoshuk's team was mainly concerned with the use of adjectival words as measures in their critique of the scaling methods used in a wide variety of fields. However, the problem is clearly not limited to just these verbal descriptors. Other researchers have used magnitude estimate techniques to assign numerical estimates to the strength of a taste experience (see Stevens, 1971b, for the classic justification of this procedure) and have observed power functions with exponents of about 1.3 for sucrose and salt respectively.

However reassuring that numbers, rather than vague adjectives, are used in magnitude estimation experiments, both means of scaling intensity are subject to the same criticism: namely, that the subjective strengths indicated by either the words or the numbers mean different things to different observers, and may mean nothing in a strictly quantitative sense where continuity, monotonicity, and homogeneity are important criteria. Indeed, they may not even mean the same thing to the same observer on different days or even in different trials! Contrary to some claims, the assignment of numbers in a ranking task is not measurement per se; rather, it is a judgment of how something feels, and reflects, at best, an extremely elastic ordinal scale, one that hardly deserves to be called quantitative. The comments of Michell (1999) are especially relevant in this context.[12]

Social psychologists (Biernat, Vescio, and Manis, 1998; Biernat and Kobrynowicz, 1997; and Biernat, Manis, and Kobrynowicz, 1997) have also shown that the same problem can occur in studies of human social psychology. They argue, justifiably, that such studies may have more profound immediate social impact far beyond the esoterica of sensory processes. For example, racial stereotypes lead observers to use significantly different scales of accomplishment and capabilities for different racial groups. Furthermore, gender stereotypes also lead to different performance scales for men and women. The classic example, according to these social psychologists, is the U.S. Navy's evaluation of women in pilot training programs. More seemed to be demanded of women than men in a manner that seems to be a direct result of shifting ordinal scales based on gender-based prejudices and stereotypes.

Another example of the way in which this kind of scaling can influence socially relevant decisions can be found in our courtrooms. It is well established that people are less able to accurately identify members of other races than of their own race. A possible explanation for this phenomenon is that we have a kind of elastic visual space to which we assign faces. Members of familiar groups are placed close to the center of this space and are relatively widely separated from each other in it; that is, the subjective interperson interval is relatively large. Therefore, faces in this space are both familiar and distinguishable. Members of other groups with which we are less familiar are clustered tightly together in a different and usually distant part of this hypothetical psychological space, and are, therefore, less familiar and less distinguishable one from another.

This is a speculative answer developed to explain this "Own Race Bias" (ORB) that confounds jurisprudence so frequently; however, it is consistent with the outline of the other scaling methods described earlier. Whatever the true explanation may be, the facts of the matter are clear-cut. Meissner and Brigham (2001) have recently published a comprehensive metareview of the literature on the ORB. Their survey of 39 research articles made it clear that:

- The ORB effect is highly reliable. This means that it shows up to some degree in most of the studies in which it has been sought.
- The ORB, somewhat surprisingly, goes in both directions. Both black and white witnesses exhibit the effect even though the black community is a minority and exposed to white faces more often, rather than vice versa. Not surprisingly, the effect is somewhat greater for white witnesses than for black.
- The ORB occurs over a wide range of memory tasks. It does not matter what method is used, the phenomenon is robust.
- The ORB is replicated when individuals are repeatedly tested and in different testing situations.
- The ORB is especially prominent in generating false alarms—i.e., identifying a suspect who is actually innocent.
- The ORB tends to produce less correct identification and more false alarms for other races than for one's own race.

Obviously, inappropriate ranking, rating, and identification methods have many socially relevant effects beyond the esoterica of the psychological laboratory. They provide an understanding, if not an explanation, of the many instances in which stereotypy and prejudice influence our decisions. For our purposes they also provide additional evidence of the discrepancies between the laws of physics and the laws of the mind. The sliding scales, the variable intervals, the false correlations, and the potential reversal of judgments, all argue that the homogeneity that graces the physical world does not exist in the mental world.

2.7 CATEGORICAL PERCEPTION

We have seen so far in this section that there is a very strong, although not universal, tendency for observers to use a relatively small number of categories ("seven plus or minus two"; "red"; "high pitch"; "more than"; etc.) for many different judgments about magnitude and quality. Although the number of categories may differ from test to test and modality to modality, it is clear that most people[13] are severely limited in their absolute identification of virtually any dimension that one could mention.

We have also seen that the number of categories used depends on both the situation and the observer. If reference stimuli or multiple dimensions are provided, the magical "seven plus or minus two" becomes a larger number, in some cases twenty or more. In a practical, multidimensional world outside the laboratory, of course, we regularly do much, much better. Our ability to recognize faces and other objects is enormous because of the many additional dimensions we use in most practical situations. Furthermore, the categories that are used are very much determined by the experience and training of the observer, particularly at young ages. Repeated studies in a variety of fields have shown that observers tend to appreciate small differences between items in the same category, but are relatively insensitive to even larger differences for those that fall outside the boundary of a category.

It must also be recognized that there are substantial differences between the senses. Vision, as we have seen, seems to have a set of eleven or so natural categories by means of which colors are named, whereas there did not seem to be any particular tones that were identified better than any others (Burns and Ward, 1978). Absolute tone identification, however, may involve even fewer "natural categories." In any case, the general finding holds; even though we are able to discriminate that two simultaneously presented stimuli are different with a high of precision, it is very difficult to absolutely identify an isolated stimulus.

Thus, there is a substantial body of evidence that suggests that the number of categories we use to identify an isolated stimulus is relatively small. Extrapolating this conclusion to social interactions suggests that many of the problems of race and gender, of stereotypy and prejudice, evolve from this basic psychophysical property: our inability to have enough categories to describe what turn out to be wide differences in human behavior or appearance. There is also considerable evidence that training and experience would go a long way in overcoming these social evils and perhaps some of the simpler examples of our limited discriminative abilities.

One of the most interesting failures of our ability to discriminate between stimulus differences, even when they are quite large, is referred to as *categorical perception*. Categorical perception refers to the general tendency for observers to group elements together into a few categories along some dimension that should be more finely discriminated. All of this evidence supports the argument that many of the dimensions of our cognitive processes do not enjoy the stability of homogeneity, isotropy, or the other properties enjoyed by physical processes and dimensions.

Categorical perception occurs in a wide variety of different behaviors. Speech perception, as first reported by Liberman, Harris, Hoffman, and Griffith (1957), is a classic example. They described how a continuum of speech sounds ranging from sounding alike (for example, "pa" at one extreme and "ba" at the other, could be used as stimuli in a discrimination experiment. A conventional prediction is that the probability of identifying either one of these two sounds would progressively change from one end of this continuum to the other, with uncertain response probabilities occurring only in the middle of the range where the two sounds were most similar. This initial speculation, therefore, suggests that there should be a gradual increase in the report of a "pa" stimulus, for example, as the speech sound becomes more and more like "pa," and a gradual decrease in the response to "ba" as it became progressively less like "ba" and more like "pa.")

Surprisingly, Liberman and his colleagues, as well as a host of other subsequent researchers, found that their observers reported either a "pa" or a "ba" over wide ranges, with a sharp cutoff between the two categories as the continuum was scanned. That is, there was an abrupt change from perceptually identifying a sound as "pa" to identifying it as "ba," although there was a continuous change in the physical properties from one sound extreme to the other. The change in the report of either sound was abrupt, as shown in Figure 2.2.

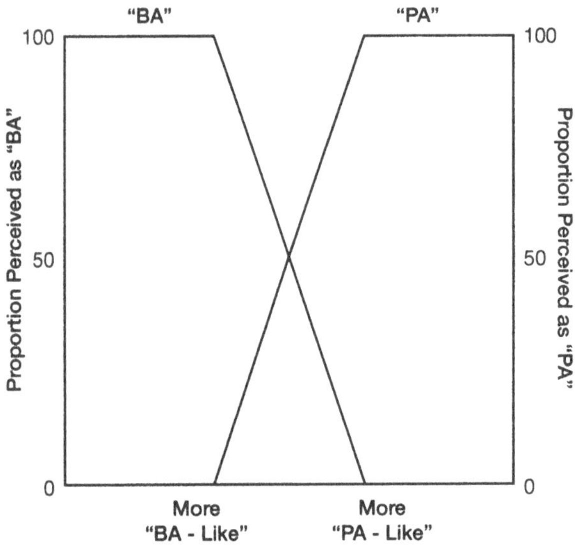

Fig. 2.2 The Classical Categorical Perception Result
Although the degree of "Pa-ness" or "Ba-ness" may change continuously, subjects persevere in their tendency to identify "Pa-like" sounds even when they are becoming more "Ba-like."

Again, if there is some additional information (such as a reference sound of "pa" provided close to an ambiguous "ba"), observers could relatively easily tell the difference between two sounds. In the absence of such a reference, this kind of absolute judgment was extremely insensitive to substantial differences between the "pa" and "ba" stimuli, respectively.

Another way in which the categorical perception effects on speech sounds can be mitigated is by the provision of additional nonvocal information, specifically motion pictures of the changing shape of the mouth as a person speaks. Massaro (1997) reported a significant body of results in which a computer-generated model of facial expressions strongly influenced and eventually overrode the categories that resulted when only speech sounds were presented. What might have been clearly a "pa" as an acoustic stimulus could be misperceived as a "ba" if the form of the lips matches that associated with the latter sound.

It is interesting to note that the categorical perception phenomenon also occurs when facial expressions alone are studied. Etcoff and Magee (1992) studied computer-generated facial expressions of different emotions. They also were able to show that there was not a smooth continuum between the perceptions of different emotional expressions; rather, their observers showed poor discrimination within a category but good discrimination between categories.

This, of course, is the defining characteristic of the categorical perception phenomenon. It works for either vision or audition alone, and each can influence the other. A good, although slightly dated, introduction to the span of topics included under the categorical perception rubric can be found in Harnad (1987).

In recent years there has been a considerable amount of attention paid to the possible neural origins of the categorical phenomena. It has been suggested that it may be due to the nature of neural feature detectors or localized regions of the brain competing for dominance. However, such fanciful neuroreductive theories run into strong counterarguments because of the experiential and learning effects that have been repeatedly shown to exist whenever this phenomenon is studied. The salient and important point in the present discussion is that the continuity of the physical stimulus is not matched by continuity of the perceived responses. Once again, a clear distinction between the properties of the physical world and the psychological world is demonstrated.

2.8 ON VALIDITY

The concepts of reliability and validity are scattered throughout the history of psychological science. *Reliability*, defined in terms of repeatability or internal consistency, is relatively simple to define. If an experiment is repeated, or if two parts of a single test are compared to each other, and the results are, within certain agreed upon limits, the same, the experiment is considered to be reliable.

Validity, however, is a much more difficult concept to define. By *validity*, we mean something much more subtle and complicated than reliability. Validity has been defined in many ways including: (1) *face validity* (a superficial examination appears to show that a measure is actually measuring what we suppose it to be measuring); (2) *criterion validity* (the degree to which our new measures agree with other results); or (3) *construct validity* (the relationship between the measure and other possible measures incorporated within the same theory, Messick, 1995). In general, a rough definition of validity is that a measure is valid to the extent to which it truly represents (i.e., measures) the properties of that which is intended to be measured.

Unfortunately, it is far more difficult to specify that a measure is valid than is generally appreciated. The problem is one of circularity. To know that a measure is "valid," we must carry out a measurement and then find some other way to link it to the property that does not depend on the measurement. Any valid measurement along any dimension, therefore, requires that we have an a priori estimate of what are the dimension's properties and some a posteriori means of linking the measurement and the dimension. Setting aside for the moment any strategies such as convergent operations or validity defined in terms of its pragmatic utility[14], it is clear that at some point in the process, a subjective and consensual estimate of validity that transcends any possible objective criterion is going to have to be made.

Therefore, validity, regardless of type, has to be understood as a judgment, an evaluation based, in the final analysis, on the "wisdom of experts." Although there are some techniques that can help to link a measure with some psychological construct less tentatively, what is obviously fallacious in "common sense" or "face validity" also holds true for the other two narrower, more technical definitions of validity. The problem is seriously exacerbated when the dimension being measured is otherwise inaccessible and likely to be causally related to many different interacting factors.

The problem of validity is closely tied to the problem of the arbitrariness of the metric used to quantify a dimension that is not directly accessible. Messick (1995), for example, proposed a strategy for thinking about the attachment of metric meaning to a measure when he discussed how one maximizes the "construct validity" of a measure. He suggested that validation depends on both empirical tests and rational argument:

> As such, validation combines scientific inquiry with rational argument to justify (or nullify) score interpretation and use. ... In principle as well as in practice, construct validity is based on an integration of any evidence that bears on the interpretation or meaning of the test scores—including content—and criterion-related evidence—which are thus subsumed as part of construct validity. (p. 742)

Other even more strained efforts to attach meaning to arbitrary metrics and establish zero points (in order to reinforce the validity of psychological measurements) have been generated by psychologists frustrated with the elusiveness of the concept. Blanton and Jaccard (2006) suggested that the following steps should be used by psychologists to "attach meaning" (i.e., validity) to what are often meaningless metrics.

a. Identify the relevant events they view as meaningful.
b. Make a case for the importance of these events and the positioning of these events on the underlying psychological dimension in an absolute sense.
c. Build consensus among members of the scientific or applied community about such positioning.
d. Conduct the necessary research to link test scores to those events in such a way to render the metric of the test meaningful.
e. Make a case and build consensus for the threshold values to make diagnostic statements. (p. 38)

None of this advice, unfortunately, seems to offer a means of rigorously establishing that a measure is really measuring what a psychologist thinks it is measuring. Blanton and Jaccard's (2006) suggestions b, c, and e sound more like a publicity campaign than a search for scientific rigor. Suggestions a and d fly in the face of the basic problem; scales that are intrinsically meaningless or invalid cannot by further research be transformed into valid ones.

Messick's (1995) strategy for attaching validity to measures also carries unavoidable overtones of *argumentation* as apposed to *proof*. His "integration" criterion is comparable to the idea that convergent results can provide precision when individual results are imprecise. Both ideas raise serious concerns that a false kind of "truth" can arise from what are invalid, or indeterminate, data. If we followed Messick's line of advice, establishing validity or making non-arbitrary that which is fundamentally arbitrary would become a popularity contest, not an example of robust scientific proof, especially for psychologists who have little direct access to their mental constructs.

The concept we should carry away from this brief discussion is that if a phenomenon is fundamentally arbitrary, no amount of argumentation, consensus building, or "rational argument" is going to remove that arbitrariness. Psychologists' efforts in this direction to confirm the validity of a diagnostic test or a theoretical concept are futile and can provide no effective remedy to overcome what is an insurmountable inaccessibility of the inner workings of the mind, no matter how high the correlation between stimulus and behavioral response.

Validity and reliability have, therefore, long been known to be special problems for cognitive science. However there are other factors of a more specific nature that are not always discussed in the context of measurement. In the following section, I deal more specifically with the requirements for measurability, specifically considering the notion of quantifiable dimensions.

2.9 FURTHER COMMENTS ON QUANTIFIABILITY

One of the primary properties of a quantitative system is that it adheres to the arithmetic rules. However, this simple and basic property is not the sole criterion to guarantee that a measurement is really susceptible to quantitative measurement. As discussed on page 28, there is an ensemble of properties that must be fulfilled for a phenomenon to be susceptible to quantitative measurement. An event or phenomenon must:

- Follow the laws of arithmetic, specifically additivity
- Possess a non-arbitrary zero and a non-arbitrary metric
- Possess a constant unit of measurement
- Exhibit equality
- Exhibit rank order, specifically ordinality
- Exhibit equality of intervals
- Exhibit equality of ratios
- Be capable of being transformed from one system to another simply by multiplying each value in one system by a single number.

Although measurements of overt behavior usually do meet these criteria, since they involve well-established physical dimensions, the covert properties of the mind seem less likely to satisfy these criteria.

Let's consider a few of the phenomena that suggest that mental processes are not only not accessible and measurable, but may also be inherently non-quantifiable.

2.9.1 On Additivity

One of the most time-honored myths of experimental psychological research is the method originally proposed by Donders (1868/1969) to study reaction times. Donders' idea was a simple one; if a reaction time was composed of a series of separable states (e.g., perception, stimulus identification, response selection, and motor response, respectively), why not design an experiment so that each component could be selectively removed to determine how much time it added to the overall reaction time?

A modern version of this method was designed by Sternberg (1969) and was designated as the *additive factors method*. Sternberg's logic was based on the idea that two properties of the stimulus would statistically "interact" if they were operating on the same "mental component," but their effects would simply add together if they were operating on different components.

Both of these methods are deeply flawed, not by virtue of any practical or methodological difficulty in applying them in an experimental protocol, but rather by the fact that their most fundamental assumption is incorrect. The flawed assumption is that the components of cognitive activity can be added or subtracted in the same way as can a number of beans or the days of the week.

There is increasing evidence that the idea of separate and independent mental components that can be added or subtracted from each other, leaving the rest of the system otherwise unchanged, is just flat out wrong. Not only is the hypothesis of independent processing stages probably incorrect, but the simplicity of how they interact is vastly overestimated. That is, the idea that these components (even if they exist) can be added or subtracted without effect on the rest of the system is not justified.

This assumption—pure insertion—which lies at the heart of so much psychological experimentation and theorizing, assumes something that is not even approximated in experiments. When a process is simplified by changing the task demands, the overall system responds not by adding or subtracting a constant effect; instead, the entire system adjusts.[15]

The suggestion that we can distinguish between "additive" and "interactive" processes is not supported by the data obtained in such experiments. The usual criterion used to make this distinction—crossed or parallel functional relationships—is rarely clear-cut. As Pachella (1974) pointed out, few experiments in psychology resolve such controversies in a definitive fashion; they simply do not have the precision to make the distinction between one hypothesis and another. The results, after much repetition, usually end up supporting some intermediate hypothesis, and irresolvable controversy continues.

The continuing debate over whether cognitive processes occur in a parallel or serial manner is another example of such a continuing, but ultimately unresolvable, controversy. Nowadays, it seems much more likely that the complex network of cognitive forces exhibits both organizational schemes simultaneously. However, even this hypothesis has to be constrained by the indeterminate nature of psychological findings concerning the extent to which arithmetic rules are followed in our thought processes.

A recent reconsideration of the problems of additivity in psychology has been offered by Michell (1999). He pointed out that additivity is not a given but is something that must be discovered by empirical tests. For Michell, as for Helmholtz, to whom he regularly refers, quantifiability depends primarily on additivity (as well as ordinality). This requirement means that additivity must be empirically determined for a mental process before we can begin the process of quantifying and thus "measuring" a psychological process. If a process can be shown to be both additive and ordinal (among the other properties listed on page 58), it is presumptively quantitative, and only then can we appropriately design a protocol to carry out the measurement. If it cannot, as most psychological dimensions cannot, then empirical quantification becomes a quest for a nonexistent chimera! However, even if its quantifiability can be established, there is still no guarantee that the outcome of the experiment will be precise enough to support a theory of one kind or another or to distinguish among competitive theories.

Unfortunately, as Michell also pointed out, empirical tests of additivity are rarely carried out. He further noted that psychologists incorrectly assume that all mental processes are a priori measurable. On reflection, it seems quite clear that a host of psychological experiments have shown that the criteria of additivity that Michell, Helmholtz, and others have identified as a requisite for quantifiability are not easily and certainly not usually met, in psychological research.

How is additivity defined formally? Michell defines it in the following way in the glossary of his important book:

> Additivity: A relation between levels of a quantitative attribute. For any two distinct levels of a quantitative attribute, a third always exists such that the greater of the two is the sum of the third and the less. These three levels are related additively. Additive relations must be commutative and associative. (p. 220)

Although quite easy to put into words like these, tests that seek to determine additivity are extremely difficult to carry out. Statisticians have devised many tests of additivity (for example, see Tukey's (1949) test for non-additivity, and Chan, Christofferson, and Stenseth's (2003) Lagrange multiplier test). But such tests are rarely, if ever, carried out in the usual psychological experiment. Indeed, the nature of the additivity problem remains controversial (see, for example, the recent encyclopedia article by Karabatsos, 2005) and remains a matter of active debate among statisticians. Yet, without additivity, the entire concept of measurement comes into doubt; the chain of logic requires that to be measured, something must be quantitative, and to be quantitative, it must be additive, among other criteria.

The suspicion, if not the proof, is that many cognitive phenomena (as well as observations from many other fields of science) are neither additive, nor quantitative, nor measurable in Michell's sense. This is a compelling argument for the inaccessibility of cognitive processes. It is also a fundamental criticism of the cognitive mentalism prevailing in contemporary psychology. Furthermore, if the additivity criterion does not hold, then, also, must all ideas of using arithmetic cum mathematics as a tool for analyzing cognitive phenomena fall by the wayside. I must reiterate, however, that this constraint does not hold for the measurement of behavior.

Additivity is also a major, but flawed, criterion for the validity of cognitive theories of mental modularity. That is, to the extent that stimuli interact and their functions do not add, any suggestive evidence supporting independent modules and "pure insertion" becomes problematic. Nevertheless, as rare as it is to find any evidence of cognitive additivity, more or less independent modularity is widely assumed by contemporary cognitive neuroscientists as a basic characteristic of cognitive processing.

Pachella's (1974) admonition that the precision of cognitive methodologies is not sufficient to warrant the assumption of additivity, even in those cases in which it is weakly suggested, is worth reiterating at this point. The high degree of interaction existing between the effects of what appear to be diverse stimuli suggests that the mind-brain system operates more holistically than it does as a collective of independent modules. It appears that the opposing opinion, that both cognitive processes and brain representations are modular is gradually changing. However, the extreme view, akin to a modern version of phrenology, still has a seductive attraction to cognitive neuroscientists.

In sum, the implication of a lack of additivity in psychological space-time is that our so-called "measurements" of complex, interactive mental processes may be less meaningful or valid than they at first seem. A demonstrated lack of additivity is one of the most basic indications that cognitive quantification is not always possible and, therefore, that conventional mathematical models may not be appropriate models of cognition.

Tests of additivity aside, the criterion of psychological ordinality also remains extremely elusive. Indeed, I dedicate an extensive section of Chapter 3 to this problem later in this book under the rubric of temporal paradoxes, another name for the non-ordinal nature of many psychological phenomena.

2.9.2 On Arbitrary Zeroes and Metrics

Are cognitive dimensions anchored by a non-arbitrary zero? This is another of the essential properties required to establish the quantifiability and, thus, the measurability of mental processes. However, almost all psychological dimensions have arbitrary zeroes, particularly when one leaves the domain of overt behavioral measurements for the less tangible domain of cognitive processes.

Similarly, we can ask: Do cognitive dimensions exhibit adequately defined metrics of their assumed dimensions? As previously defined, a metric is a "geometric function that describes the distances between pairs of points in a space." Given that such a function is required for orderly quantification, and that such metrics are elusive, at best, in psychological measurements, further questions have to be raised about the quantifiability of cognitive processes. These questions are so fundamental that it is desirable to expand on the introductory comments made in Chapter 1.

One of the most interesting and relevant articles considering this issue was recently published by Blanton and Jaccard (2006). They discussed the problem of arbitrary metrics and zeroes in psychology and pointed out that most conventional psychological processes are "bipolar" measures with "neutral midpoints." They noted that these midpoints are often assumed to represent a "zero" value for the psychological process under study. According to Blanton and Jaccard, "such assumptions are not warranted … [and are] assumed to be true by fiat" (p. 33). They concluded that we have to go far further than the usual criteria of validity and reliability to accept psychological measurements as scientifically sound measures. They thus added the nature of the metric and the scale to the criteria for quantifiability and thus measurability. Furthermore, they reminded us of the necessity for a non-arbitrary zero point on that scale.

Since one defining feature of a quantifiable metric (i.e., of a ratio scale) is its non-arbitrary zero, the implication of their discussion[16] is that most, if not all, efforts to measure psychological hypothetical constructs are less than fully quantifiable. Their important contribution was to remind psychology of the frailty of its system of arbitrary metrics whenever we move beyond the boundaries of publicly observable behavior.

An important related point is that zero values are especially arbitrary in psychology, just because they are so modifiable by cognitive processes of which we know very little. Thus, when we measure something with an interval scale we will always be uncertain exactly what is being measured. This is another way of expressing the uncertainty expressed by the "one to many" caveat discussed on page 39.

This also underscores the fact that although interval scales are widely used in psychology to "measure" such properties as hue or image quality, they are inadequately quantifiable and can lead to bizarre conclusions. Stevens (1975) understood this problem when he referred to "prothetic" and "metathetic" dimensions, respectively. *Prothetic dimensions* (such as loudness) are the potentially quantifiable ones that are, in some sense of the word, additive; this is the stuff of a quantitative science. *Metathetic dimensions* are the qualitative ones (such as hue) for which interval scales may, in the absence of real quantifiability, be used with care, but which do not meet the conditions for robust quantifiability. Measurement, however, depends on the availability of a prothetic dimension that meets all of the other conditions for quantification previously tabulated.

Despite this emerging awareness of the necessity of non-arbitrary zero points, robust metrics, and quantifiability for any scientific analysis, psychological research designs have incautiously accepted a variety of scales for studying human mentation and behavior. Although it now seems clear that only ratio scales provide a robust foundation for quantification, nominal, ordinal, and interval scales are often used to measure mental activity, and the results obtained used as arguments for one kind of a theoretical explanation or another. Unfortunately, any data obtained with these less robust scales are indeterminate and inadequately exclusive with regard to the exact underlying cognitive mechanisms. Thus, it is possible to generate an infinite number of explanations from them, many of which are equally plausible.

If the zero point is arbitrary, what can be said about the function that defines the separation of distances between adjacent psychological unit values: the metric? The answer to this rhetorical question is that the intervals are also likely to be arbitrary. This does not just mean that they are unequal. Instead, it implies that they are often irregular (with no obvious metric) and cannot be dealt with in the manner that permits us to appreciate the meaning or significance of a change in the measured value. Blanton and Jaccard (2006) make the point quite eloquently:

> We define a metric as arbitrary when it is not known where a given score locates an individual on the underlying dimension or how a one unit change on the observed score reflects the magnitude of change on the underlying dimension. (p. 28)

For a system in which the "underlying dimension" is inaccessible, it becomes extremely difficult, if not impossible, to evaluate the mathematical expression defining the separation (i.e., distances) between successive points. This adds a further complication to the lack of meaningfulness (see page 24) of a metric with an arbitrary zero.

Yet, this issue has also been generally ignored by psychologists, who blithely plunge ahead to make very specific associations between the behavioral measures obtained in an experiment and the underlying cognitive dimension. Consensual agreement on what a metric might be is not the same thing as an empirical determination of its form. Unfortunately, if the underlying metric is inaccessible, then no such empirical determination is possible.

In such a situation, the hope of establishing the quantifiability of these underlying cognitive dimensions in a robust fashion becomes increasingly remote. What psychologists have substituted for rigorous quantifiability is a kind of squishy acceptance of verbal or correlative description which is, at best, among the weakest forms of explanation.[17]

Herein lies the very essence of the source of some of the most frustrating of psychology's scientific difficulties: its endless number of unconfirmable, redundant microtheories. The result is a lack of convergent pyramiding in which a few general rules can be used to explain a significant body of empirical findings. The ultimate cause is the immeasurability of inaccessible mental processes, a deficiency that arises from the absence of the necessary properties that define quantifiability.

2.10 INTERIM SUMMARY

This chapter has expanded some of the most basic attributes of psychological quantifiability and measurability and presented a few examples of its failure in human performance. It highlights a number of instances in which the necessary conditions for quantifiability are not exhibited by cognitive processes. Our powers of dealing with numerousness are highly limited, not only in animals, infants, and adults from undeveloped cultures, but also in normally functioning adults. Furthermore, we seem to have a poor ability to apprehend numerousness without explicit counting. Nor do the units of psychological space seem, in general, to be evenly spaced or equal in size. Whereas quantification is a necessary aspect of all of the dimensions encountered in the physical sciences, the properties of the human mind seem to violate the requirements for orderliness, regularity, and ratio scaling. In failing to exhibit these properties, they fail to meet the basic standards for quantifiability and thus measurability. The isomorphism between numbers and events (or objects) that Stevens (1951, p.23) depended on just does not hold for psychological phenomena in the same manner it does for physical ones.

Furthermore, people encounter profound difficulties in responding differently to stimuli that may be quite different from each other, not only in their appearance but also in their origins, unless we provide additional information in the form of a comparison stimulus or other simultaneous dimensions. This weak discriminative ability of both quantity and quality is also strong evidence that our ability to quantify the cognitive results of physical stimulation is relatively poor compared to a physicist's ability to measure physical dimensions.

Although it cannot be definitively proven because of the general barrier of inaccessibility, it seems likely that the dimensions of cognitive experience are highly irregular and do not have a simple metric. A just detectable difference means something quite different at one end of a stimulus-intensity spectrum than at another (e.g., Weber's Law), and there are other instances (e.g., visual wavelength discrimination) in which they are non-uniform.

These problems, I must emphasize, do not mean that the nominal, interval, or ordinal scales[18] cannot provide useful information in many cases. However, they do mean that the measures made with these scales are indeterminate and do not provide enough information to infer the properties of the covert cognitive mechanisms and processes that encode them.

Nowhere is this difference between interval and ratio scales clearer than in the scales used to measure temperatures. Both Centigrade and Fahrenheit thermometers are based on interval scales. They permit the collection of useful data about the behavior of our bodies and our environment. However, because they are based on an arbitrary zero point, they cannot be used reductively; that is, these measures do not help to identify the underlying physical processes of heat. It was only with the emergence of new concepts that became possible with the Kelvin scale (with its non-arbitrary zero) that modern appreciation of what temperature actually means (heat is motion) emerged.

The properties of mental activity examined in this chapter suggest, therefore, that the mind, in general, does not exhibit the properties necessary for robust quantification. Psychologists have been much too content to deal with some very loose "measures." For example, many of the scoring procedures for various kinds of diagnostic tests, surveys, and even some experimental results exhibit only the weakest form of rank-order scaling. Whereas strict quantification requires that ratio scales be used, many of our measures are actually degraded forms of much less meaningful interval or ordinal scales. Unfortunately, the underdetermination that results from the use of scales opens the door to inadequately precise theoretical explanations; that is, none of the theories can be discriminated from each other. Instead, almost any theory can be made to seem at least plausible, if not possible.

Perhaps it is in the negative answer to the basic question: Are inaccessible psychological constructs quantifiable? that the explanation for the lack of pyramiding and the proliferation of indistinguishable psychological theories is to be found. Contemporary psychological theories are all too often narrowly conceptualized, experiment specific, and isolated from each other. Empirical tests, the basic material of improving the strength of an explanation or theory, are too loosely linked to the putative explanation. In such a context, the enormous variability of human behavior provides ample opportunity for discovering "supporting evidence" to satisfy almost any "plausible" theory.[19]

Psychology has, without much explicit discussion, attempted to overcome these handicaps by seeking alternative strategies to remove arbitrariness and indeterminativeness and to enhance the rigor of psychological theories. However, most of the strategies that purport to do so are actually calls for the development of a subjective consensus by experts based on very fragile foundations: the accumulation of additional, though equally flawed, empirical observations or some kind of non-quantitative argumentation to establish the primacy of one putative theory over another. Because of the impotence of this approach (due entirely to fundamental, in principle barriers), psychological explanations remain highly specific and fractionated.

The current state of much of contemporary psychology can, therefore, be summed up as follows:

1. Measurements that are based on nominal, ordinal, or interval scales are indeterminate with regard to the underlying dimensions. For data accumulated under these restricted conditions, it is impossible to remove the arbitrariness of neutral points and establish robust dimensional metrics.
2. Therefore, there can be no robust guarantee that the laws describing mental processing are the same as those describing the physical stimulus world. We do not have the justification for inference and extrapolation that is provided to the physical sciences by the Cosmological Principle.
3. Furthermore, even when ratio scales are available, there are other reasons (including the one to many rule) to suggest that behavioral findings are incapable of uniquely defining the nature of the underlying cognitive mechanisms. Bridges between behavior and mentation are fragile and incomplete because of limited quantifiability and indeterminativeness of mental phenomena. This means that most psychological experiments carried out to distinguish between two plausible theories rarely have the power to make such a discrimination. To depend on such secondary criteria as Ockham's razor or some ill-defined concept of elegance in such a situation is a recipe for misdirection. Although an enormous effort is made in psychology graduate education programs to teach methodology, little attention is paid to the concepts underlying the ability of methodology to make meaningful decisions about putative explanations.[20]
4. Physical phenomena and metrics, on the other hand, are regular and generally meet the other criteria discussed here for quantification and measurement. In those cases where irregularity occurs, there are often explanations based on axiomatic-deductive procedures.[21] Common laws and explanations linking many different physical phenomena exist in all but the most extreme cases, most notably the residual chasm between quantum and relativity theories. Psychological phenomena and metrics, on the other hand, are generally irregular and do not usually meet the criteria for quantification and measurement. Psychological processes are not linked by common laws and are seemingly independent of each other. There are, for all practical and theoretical purposes, no common theories of the phenomena discussed in this and the next chapter.

If this line of thought is correct, it means that many of the widely accepted theories and explanations that flood psychological science are misleading if not incorrect. Stevens' (1951) effort to justify the use of scales other than the ratio scale as suitable tools for psychological research is, of course, the classic example. However, as scholars such as Michell (1999) have illuminated, although there is no difficulty making such measurements, they actually produce very little information that can be used to infer underlying processes, and can lead to serious distortions in our understanding of the most basic properties of cognition. Although rating and interval scales may easily lead to neat looking and orderly graphs, closer examination reveals serious disorder and even reversals of rankings when different observers are compared.

Stevens' attempt to authenticate what are essentially non-quantitative dimensions (i.e., those based on nominal, ordinal, and interval scales) as valid metrics of psychological function led to fundamentally flawed psychophysical strategies. Not only does the use of these scales prevent us from inferring underlying processes and mechanisms, but the raw data may distort reality. Although "measurements" may be made and "numbers assigned," those numbers may not convey even the simplest meaning. The problem was eloquently summarized by Michell (1999):

> if Stevens' definition of measurement [as simple assignment of numerals] is accepted, then the scientific task of quantification is cancelled and only the instrumental task remains. The scientific task is cancelled because Stevens' definition is indifferent to the structure of the world. His definition requires no quantitative structures. It eliminates the scientific task and leaches the instrumental task of its scientific content. (p. 77)

One implication of this train of thought is that most of psychology's empirical results are indeterminate with regard to inner processes and mechanisms. The more serious implication of this state of affairs, however, is that it means that psychology cannot *in principle* develop broadly integrative theories. All of the fractionation and all of the microtheories that characterize our science these days are not results of inadequate intellectual tools, but rather are the results of a fundamental constraint on the quantification of psychological phenomena. This logic can be summed up as follows:

1. Mental dimensions do not in general exhibit the properties of ratio scalability.
2. Therefore, mental dimensions are inherently unquantifiable.
3. Therefore, valid measurements of mental processes are not possible.
4. Therefore, the "measurements" that are made are indeterminate with regard to underlying mental processes and mechanisms.
5. Therefore, inference from behavior to mental processes and mechanisms is not possible.
6. Furthermore, since such measurements are indeterminate, no other conventional scientific tools such as mathematics can be applied in a way that offers

> the hope of reductive explanation. At best, mathematical models are descriptions of behavior.

However, it must not be forgotten that there is a scientific mode in which psychology can flourish. It is an objective, quantifiable behaviorism, a scientific approach that deals with response dimensions that are inherently ratio scalable.

NOTES

[1]Because of some possible ambiguity in the meaning of the term *Cosmological Principle*, it may be well to define it specifically once again here. This important physical statement means that there are no special places in the universe; no matter what direction we look, the properties of space and time are the same. Specifically, if we examine a large enough space, the universe is homogenous and isotropic. This means that the distances between adjacent points in space are constant. A vital corollary or implication of this principle is that the laws of physics are the same everywhere.

[2]The crux of the "one-to-many" rule is that when the mechanisms of some system are not open to direct observation (i.e., they are inaccessible), there is a plethora of possible and plausible explanations, each of which can equally well explain a particular behavioral observation. This argument continues by asserting that there is in general no possible empirical strategy to distinguish among them. The application of this principle to psychology is a specialization of the same idea expressed in many other fields of science, including automata theory. One of the most explicit expressions of this fundamental truth was the second theorem offered by Moore (1956) in his classic article on the analysis of closed systems. It is also implicit in the "black box" constraint well known to system engineers.

[3]The reality of mind is, to an unfortunate degree, still a debatable issue. The problem is that there is only one direct piece of evidence that our minds are real: the awareness each of us has of our own sentience. Distinguishing between a conscious brain and an unconscious automaton is a terribly difficult task. No strategy has yet been offered that would allow us to do so. Although we cannot predict the future, it may be that this controversy will remain with us for quite a while.

[4]The problem of how our brain represents numbers is a source of considerable debate. Spelke and Dehaene (1999) and Simon (1999) have argued that there are (or are not), respectively, specialized brain circuits accounting for our mathematical abilities. Others have joined in the debate by arguing that "no brain area is specific for subitizing and counting" (Piazza, Mochelli, Butterworth, and Price, 2002, 444). For reasons I have discussed in detail (Uttal, 2001), I do not believe it possible to resolve this argument about brain loci and prefer to concentrate on the behavioral evidence.

[5]When I say "natural" in this context, I really mean "normal." A few otherwise severely intellectually handicapped people are capable of extraordinary numerical skills. These individuals, known as "savants," are exceptionally rare but often display arithmetic or mathematical competence that vastly exceeds the normal. Although it has been suggested that a savant's exceptional mathematical skill may be the result of a release of skills we all have (Snyder and Mitchell, 1999), very little is really known about the origins of this exceptional behavior. An interesting case history, including brain scan information, has recently been published for a well-known savant, Kip Peek, by Treffert and Christensen (2005). A curious but unexplained anatomic aspect of Peek's brain was that the corpus callosum appeared to be almost entirely missing. Unfortunately, savantism remains one of the many inexplicable phenomena of the mind-brain. We can only speculate how this combination of abilities and disabilities emerges.

[6]Miller and most of the authors cited in this paper used the information score (in bits) to present their data. Transmitted information measured in bits (I_b) of an event (E) is defined as: $\log_2$ p (E), where p (E) is the probability of E occurring from among all of the possible events. The value of this measure is that it was not limited to integer values and it could be applied to a wide range of dimensions and measures. Thus, it provided a universal means of comparing performance in different tasks. For simplicity's sake, I have used the approximate integer values to make the point that human beings have very modest abilities to deal with numerousness, not what the specific value is.

[7]I am grateful to Professor Diana Deutsch of the University of California at San Diego for advice on the topic of absolute pitch as I prepared this section.

[8]In fact, Weber was neither the first to express this law, nor did he ever put it into this mathematical form. Mathematicians including Euler and astronomers such as Steinheil and Pogson had previously noted the relation for musical tones and stellar magnitudes, respectively. Fechner (1860/1966) was not only aware of this historical fact and was the one who formalized it into this famous approximation, but was also the first to show that "Weber's Law" was a general, but at best an approximate, law of perceptual processing.

[9]I use the words "color" and "pitch" here as convenient artifices. However, it should be appreciated that we know nothing directly about these perceptual experiences. Rather, all of our experiments examine the behavioral effects produced by photic wavelength and acoustic frequency: the respective physical stimulus dimensions. I hope my readers will accept this shorthand designation as I present the case that there are great discrepancies between the physical and cognitive worlds.

[10]This essentially means that the perceptual experience is not defined uniquely by the wavelength of light. Quite different wavelength combinations can produce indiscriminable perceptual experiences. This is another example that the dimensions of the physical world are disconnected from those of the psychological response.

[11]Some of this section has been adapted and updated from material previously discussed in Uttal (2003). It is used with the permission of the publisher, Lawrence Erlbaum Associates, Inc.

[12]I return to consider this topic in greater detail in Chapter 4.

[13]However, there are substantial differences among people, and a few are gifted with extraordinary skill and can make use of very fine categorizations. See the discussion on page 49.

[14]Both of these strategies are extremely susceptible to mistaken conclusions. Convergent operations are always polluted by the same weaknesses as any inductive argument. There is always the possibility that some subsequent operation will produce contradictory results. Pragmatic or utility "validity" can mislead and deceive because of non-objective factors that border on the anecdotal. In such a situation, the utilitarian needs can dominate the objective reality, to the point of the total corruption of the concept of validity. For example, a drug may seem to work to cure a disease when, in actual fact, it may have done nothing beyond that done by a placebo or even just waiting.

[15]I am indebted to the prescient insights of my colleague Robert Pachella (1974) of the University of Michigan for clarifying the frailty of this oversimplistic idea of simple additivity or pure insertion in psychology.

[16]Blanton and Jaccard (2006) went on, in their interesting article, to suggest ways in which arbitrary metrics might be used successfully in psychology. This assertion was based on "the hope is that they provide sufficient information to test psychological theories" (p. 27). They are not as negative as I am concerning the barriers to understanding imposed by arbitrary zeros. Nevertheless, their critique is vitally important in setting standards for psychology.

[17]The limits of the correlative approach are further developed in Chapter 4.

[18]Even simple nominal scales, which make no pretense of quantification, can be useful in some cases. At a minimum, they allow us to identify objects as different from each other. We would be in

sad shape if we could not name people in a way that determines that Sam is not Jeff. Unfortunately, these names tell us nothing about Sam's or Jeff's properties.

[19]The concluding line in many psychological articles has always been of interest to me. Qualifying statements such as "this evidence is not inconsistent with my theory," or "this evidence suggests," or "more research is required to resolve this issue" are indicative of a loose connection between observation and interpretations. The fact that so many of our disagreements remain unresolved either reflects the complexity of our science or that many of the methods we use are actually inadequate for the task.

[20]It may be appropriate to point out again that the inadequacy of psychological experiments to overcome inaccessibility does not mean that the behavioral findings themselves are without merit. As an example, Human Factors research (or its equivalent, Engineering Psychology), which is, in the main, atheoretical (it emphasizes the relationships between the stimuli and the elicited behavior without a lot of theoretical hand waving) is as good descriptive science as any found in the domain of physics. Human factors research, as it is pursued these days, is essentially a pure behaviorism that eschews almost all of the reductive theorizing typical of so much of cognitive psychology. Furthermore, almost all of the *empirical findings* obtained in any subdivision of psychology are, within certain limits, pretty good descriptions of behavior. The difficulties merge when we attempt to infer something about the cognitive processes and mechanisms. It is there that the frailty of the bridge between observation and behavior becomes most obvious. All such reductive theories, mathematical models included, are, in principle, incapable of providing uniquely justifiable inferences about inner processes and mechanisms.

[21]So, too, are comparable explanations often present in the sensory and motor systems. My argument here is that such reductive theories are almost totally missing when we deal with high-level cognitive processes.

3

Psychological Paradoxes in Time and Space

3.1 INTRODUCTION

In the previous chapter, I examined the difficulties engendered for measurement in cognitive research because of the arbitrariness of the mental dimensions, the poor psychological approximations to cardinality, and unequal intervals among a number of the other properties that are necessary to collectively define whether or not a dimension is quantifiable.

In this chapter, I turn to another group of the basic properties of time and space that have made it possible for physics to succeed in drawing inferences about inaccessible events and objects, whereas psychology cannot. In trying to study psychological time and space, we are confronted with processes that are not uniform; in which these dimensions are stretched and distorted; and situations that can even occur out of temporal order. This belies the hopes of those who assume that cognitive processes are, like equal-interval physical time, inelastic and monotonic. A main argument of this chapter is that the smooth monotonic flow of time is not a demonstrable property of our mental world.

Before entering into detailed discussion of how some psychological processes seem to violate properties of sequence and geometry, it is useful to precisely define my terminology. I start off with the term *inference* because it is central to everything else we discuss. An inference is defined by Webster's *Third International Unabridged Dictionary* as:

> The act of passing from one or more propositions, statements, or judgments considered as true to another, the truth of which is believed to follow from that of the former. (p. 1158)

Mathematical deduction is the strongest form of inference because of its strict rules for going from one "proposition" to another. In mathematics, the first set of propositions is the set of axioms and the second is the set of deduced theorems. In psychology, the observed data (i.e., behavior) represent the first set of propositions and the second includes the hypothetical constructions of the nature of the mind. In either case, it is necessary to establish that the "truth" of the second "follow[s] from that of the former." In mathematics, this is assumed if the deduced theorems fit the empirical facts of subsequent observations.

In psychology, however, there is no simple proof of the truth of an inference because the inference is not confirmable by direct observation. All that can be done is to consider one of many plausible and possible hypothetical outcomes. Why should this be the case? I dealt in the previous chapter with one reason: the lack of arithmetic quantifiability. I now turn to another set of reasons that collectively indicate that the most basic properties of time—ordinality, continuity, monotonicity, and sequential causality—necessary for physical inference are not evident in psychological phenomena. Having established my terminology, I will then selectively review the empirical literature to show the widespread nonveridicality of temporal and spatial phenomena from the physical stimuli that define them. These nonveridicalities or illusions attest to the fact that there is no psychological analog of the physicist's Cosmological Principle and, therefore, inferring mental states from behavioral observations is a much more difficult, if not unachievable, challenge than that faced by physicists.

To fully appreciate the argument that I am making here, it is important to be more precise about the meaning of several key words and ideas than I was in Chapter 1. The most germane to psychology include:

1. *Ordinality* refers to the generally accepted notion of a fixed sequentiality. That is, things or events are said to be ordered along some dimension in such a way that they retain their order no matter how the dimension may be transformed. For example, in the physical world, a temporal series of events that is determined to come "first," "second," or "third" maintain that order regardless of any change in the units of the dimensions. Furthermore, it is not possible for something that came "third" to occur earlier than something that came "second" by any linear transformation. To do so would violate what we mean by physical ordinality.

2. Continuity is defined in mathematics in a very specific manner. A function is said to be continuous for $x = a$, if $\lim f(x) = f(a)$ as $x \to a$. However, it is suffi-

cient to use a less formal definition. Namely, a function is continuous if nearby points are within some arbitrarily small distance from each other. That is, a function is discontinuous if adjacent values abruptly change. Sudden discontinuities occur in both physics and in psychology, of course (a sudden fracture in an earthquake, on the one hand, and a reversible image, on the other). Physical time, however, is considered to be uniformly continuous; there are many instances, as we see in this chapter, in which psychological time is not.

3. Monotonicity is closely related to ordinality and continuity but adds another constraint. A function is monotonic if its first derivative never changes sign; that is, if the function never reverses direction. It can maintain an ongoing increase or decrease or not change, but a monotonic function cannot change from an increasing function to a decreasing one or vice versa.

4. Sequential causality refers to the general idea that a cause must precede an effect. Putting aside such issues as feedback (where the cause may be modified by the effect) and philosophical arguments (on the difference between strong causality and weak causality), I am here referring to what is called linear causality in physics. This property is often expressed as: "Every effect must have its antecedent and proximate cause." This means that in the physical world, everything that happens must occur as a result of a preceding event, and that event must be close enough in space to exert its influence without requiring it to exceed the speed of light. Implicit in this definition is that there is a continuous, monotonic, and orderly flow of time. Otherwise, the meaning of "antecedent" and "preceding" would become nonsensical. Relativistic physics may seem to violate the concept of linear causality because of the compression of time at high speeds.[1] However, even relativistic causes are constrained to exert their influence on the future, as expressed in what is called their light cone.[2]

These definitions of sequential events and the smooth and even flow of time help us to understand some of the most basic properties of physical time and space. The task now set for this chapter is to show how the ordinal nature of physical time, in particular, is often violated in mental processing. Of course, we do not know exactly what the mind is doing; all we have to work with are verbal reports or other kinds of observed behavior. However, there is a sufficient amount of consistency that, even though we cannot determine the exact nature of the mechanisms that account for the perceived phenomena, we can determine when the smooth flow of time or space is violated. This evidence appears in situations in which the verbal reports and the stimulus order are nonveridical. As shown, there is ample evidence of this nonveridicality through a wide swath of psychological experimentation. However, the clearest and most definitive examples are to be found in studies of perception, and it is from this domain that most of the cited examples are drawn.

3.2 PARADOXES OF MOVEMENT

Among the most compelling arguments that psychological time does not follow the laws of sequential causality are paradoxical effects (i.e., time reversed) in which an observer apparently perceives the outcome of an event before the causes are perceived. Among the most perplexing of these phenomena is one of the best known of all perceptual illusions: apparent movement.

3.2.1 Apparent Motion

Apparent motion is a term that denotes several different kinds of subjective responses to temporally changing stimuli. The original observation of the phenomenon is attributed to Exner (1875). However, Wertheimer (1912) is usually considered to be the first systematic study of apparent motion. Wertheimer's article is, furthermore, often identified as the germinal document in the development of the modern Gestalt school of perceptual psychology. Gestalt psychologists identified four different kinds of apparent motion:

1. *Alpha Movement*: The apparent continuous expansion or contraction of an object when it was successively presented in two different-sized versions.
2. *Beta Movement:* The apparent movement of an object between two positions in which it is successively presented.
3. *Gamma Movement:* The apparent continuous expansion or contraction of a single object when the luminance is changed.
4. *Delta Movement*: Apparent movement in the direction from the second to the first position in which it is successively presented. Delta movement only occurs when the second stimulus is much more luminous than the first.

Of these four, Delta Movement is not paradoxical and does not illustrate the point being made here. The perceptual effects of size and luminosity are less crisp than the classic Beta Movement, the phenomenon on which I concentrate in the following discussion.

Consider the following stimulus paradigm: Two point sources of light are successively illuminated. If the spatial separation and the temporal interval between the two are satisfactory (optimally less than 6 degrees of visual angle and from 150 to 450 msec depending on the separation), the perceptual impression is of a light moving from the first to the second light. If the two lights are cycled repeatedly, the perception is of a single light moving back and forth between the positions of the two lights. Under the simplest conditions, the trajectory of the light is a perceptually constructed straight line that is the shortest path between the two lights.

This phenomenon is so robust that its behavior was formulated into a set of descriptive laws many years ago by Korte (1915). Although Korte's laws have subsequently been shown to be gross approximations by such researchers as Kolers

(1972), they give the flavor of how robust the perceptual experience can be. Korte's laws are usually expressed as the effect of the interactions between:

a. The distance between the stimuli,
b. The intensities of the two stimuli,
c. The exposure times of the two stimuli, and
d. The interval between them.

A sample of Korte's approximate laws states: "If you hold the intensity and the duration of the two lights constant, the spatial separation at which apparent motion occurs increases as the temporal interval between the two stimuli increases, but only up to certain limits." Clearly, this is, at best, a rough approximation, and far more precise determinations of the apparent motion phenomenon were provided by experiments conducted in later parts of the twentieth century (e.g., Caelli, Hoffman, and Lindman, 1978).

A major problem for psychology that emerges from this simple demonstration of apparent motion is that the causal forces seem to run backwards in perceptual time. That is, the direction of movement (determined by the second light) is perceived before the second light itself is perceived.

However one might go about explaining such a curious result, the apparent motion phenomenon is indisputatively temporally paradoxical, as first pointed out by Kauffman (1974). A curious property of apparent motion is that some information—the position of the second light—can be extracted from the stimulus before it is "seen." Obviously, we are dealing with some tricky language here when we use words like "seen" and "information," but once experiencing the phenomenon, it is clear what is happening.

The theoretical problem emerging when we closely examine the apparent motion phenomenon is that the temporal sequence of the physical stimulus is not matched by the perceived sequence. Thus, there is a nonveridical relationship between physical time and perceptual time. This disconnect, as well as the partial extraction of information prior to the perception of that information, suggests that the rules of temporal order, of linear causality, are being violated in this case by the perceptual system. Some process or adjustment, of which we know nothing, is reconstructing the temporal order of the physical stimulus into a new sequence of perceived events.

Apparent Beta motion, with its paradoxical time reversal, is obviously one example of how physical time and perceptual time do not covary. The best we can do is to attribute this process to a general property of the nervous system to "cognitively penetrate" the forces exerted by even simple physical stimuli. This is the essence of all explanations of this phenomenon, and it is inadequate; it is merely a restatement of the results and the evocation of a mysterious, inaccessible, and untestable process that adds nothing to our understanding of why the paradox occurs.

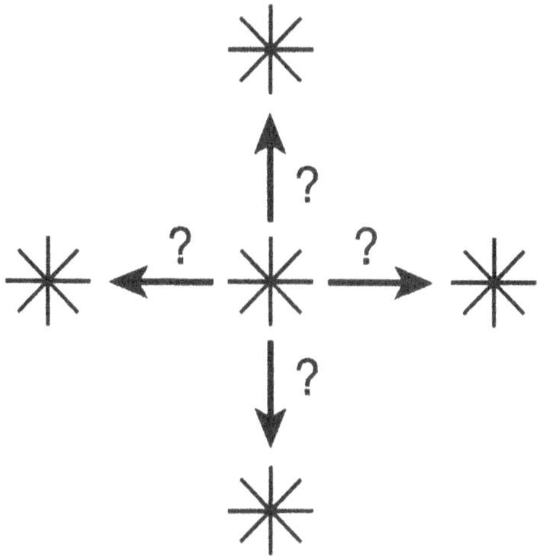

Fig. 3.1
An apparent motion stimulus in which the direction of an apparent motion cannot be determined until one of the flanking lights is illuminated. However, the flanking light is not perceived until after the apparent motion has been perceived.

The problem raised for any simple causal explanation of such profoundly disturbing phenomena as Beta apparent motion becomes clearer if we consider other examples of intrinsic nonveridicality. For example, consider the group of five lights shown in Figure 3.1. One of the lights (the one that is turned on first) is centered. Two others are positioned immediately above and below, and two others on either side. All four of these flanking lights are equally spaced and equally as bright as the center light. Uncertainty is introduced into the observer's task by using a random number to select which of the four flanking lights is turned on after the illumination of the center light. Depending upon which flanking light is illuminated, the Beta-type apparent movement experience is the appearance of the center light, motion in the direction of the selected flanking light, and, finally, the appearance of the selected flanking light.

Since which of the flanking lights would be selected was uncertain at the time the center light was illuminated, the perceived direction of motion could not be determined until the information in the second light's position was processed. Yet, there is no awareness of the second light until the end of the perceived trajectory.

Thus, the presumed causal factor (which flanking light is illuminated) in determining the direction of the apparent motion seems to have occurred prior to the perceptual awareness of that flanking light. Some information must have been available. The direction of the apparent motion was determined; that is, the observer was able to determine which one of the four flanking lights came on before it could be seen! Although there are many possible hypothetical (neural and cognitively reconstructive) explanations of this phenomenon, speaking purely from the point of view of the perceptual phenomenology, this demonstration appears to be both paradoxical and a violation of linear causality.

There are other mysterious features of apparent motion. Since the time of Brown and Voth (1937), it has been conclusively shown that apparent motion does not always follow the expected linear trajectory defined by the simple geometry of the stimulus lights. They showed that when four lights positioned on the vertices of a square were rotated around the center of that square, under certain conditions of speed of rotation and repetition rates of the lights, the apparent motion followed a curved path like that of the physical motion. The lights did not jump from light to light linearly, as would have been expected if one simply extrapolated from the simple linear Beta light phenomenon just described. Rather, the apparent motion actually curved to match the curved path of the rotating lights.

There are many other demonstrations of similar kinds of effects in which the shape of the trajectory of the apparent motion was modified by the stimulus and its environment. Depending on the timing of a set of five lights, the apparent motion can be distorted to follow a V-shaped trajectory (Antstis and Ramachandran, 1986).

In addition, the shape of the stimulus lights can have strong effects on the perceived trajectory. McBeath and Shepard (1989), for example, showed that if two images of the same object are presented at different orientations, the apparent trajectory is a curved pathway. This is another example of the cognitive penetration of the perceptual responses beyond that suggested by the raw physical geometry of the stimulus. The response is not only stimulus determined, but also influenced by a "logical" analysis suggesting the path that a physical object would have to take to rotate between the two positions.

This discrepancy between the expected linear motion and a perceived curvature can be even more extreme. If the second stimulus is not one, but two simultaneous lights at different positions, the perceptual path can easily split into two (Kolers, 1972). This interesting effect suggests that another one of the basic laws of physics—the conservation of energy and mass—is violated in some mysterious way. The mental processing seems to act as a means of preserving the logical sequence if not the physics of the event. Once again, we see a discrepancy between the laws of physics and the laws of cognitive processing.

Apparent motion can also be detoured to move along trajectories other than the simple ones specified by the two stimulus lights of the basic phenomenon. These alternate pathways can be regulated by dim gray curved pathways on the field. Although the curved paths may be longer than the shortest linear path, Shepard and Zare (1983) showed that the perceptual system responded to these subtle cues to take the longer, "more energetic pathway."

This phenomenon is especially interesting because it demonstrates that the apparent motion phenomenon is not produced by a simple spread of neural responses in a passive manner. Instead, it makes it clear that these perceptual constructions are defined by high-level, active, and "reasonable" interpretations and reconstructions. No simplistic neural theory could possibly explain these phenomena in which reasonableness and logic overwhelm physical causality. The level of complexity that is required to understand how these cognitive processes arise is well beyond any conceivable neuroreductionist approach. This is the kind of result that remains within the domain of descriptive psychology and is not subject to any "eliminative" neuroreductionism.

An even more compelling demonstration that apparent motion is regulated by logical reconstructions, rather than the defined geometry of the scene, is the finding that apparent motion can detour around objects or pictures of objects placed in what would be the usual shortest path between the two sequential lights. Meyer and Shipley (2003), for example, have shown a tendency of apparent motion to avoid passing through intervening objects. They reported that apparent motion followed a curved path around objects at relatively low curvatures. The phenomenon occurred most strongly for circular objects. All of these phenomena demonstrate the ability of an apparently moving object (a constructed phenomenon itself) to behave in a manner that is dictated by the meaning or significance of an apparently impenetrable intervening object, rather than by a simple linear transition between the two endpoints.

This kind of detour behavior can become even more complex when scene properties that seem at first glance to be unrelated to the time and distance rules formulated by Korte are introduced. Shiffrar and Freyd (1990), for example, showed at relatively long intervals (up to 750 msec) the apparent motions of parts of the body (the legs and the hands) were in directions and along trajectories determined by the anatomical constraints of the human body. For example, at moderately long intervals, an arm did not perceptually pass through the body if shown successively at two different positions on either side of the body. At shorter intervals, as short as 150 msec, however, the apparent motion followed the shortest, most direct path right through the intervening parts of the body.

Other perplexities concerning apparent motion arise as we dig deeper in the empirical research on the phenomenon. Kolers (1964) pointed out there were substantial differences between the perception of real motion and that of apparent motion. Small spots of light placed adjacent to the trajectory could be effectively masked by real movement, but not by apparent motion. He also concluded from these discrepancies that the perceptual system does not operate on the basis of any physical forces, attractions, or fields, but acts to "resolve or rationalize the disparity between two properly timed flashes" (Kolers, 1972, p.194). There have been few clearer statements of the idea that mental (perceptual) phenomena do not follow the laws of physics but are influenced by a different set of logical and structural relationships. Nor, for that matter, have there been many instances in which the implications of such wisdom have been so completely ignored.

Obviously, the nature of the two stimuli used to induce the apparent motion and the intervening visual environment are contributing factors in determining the trajectory of that motion. If, however, apparent motion was a simple process that depended only on the sequence of the lights, we would expect the trajectory to simply pass linearly from the first to the second light, no matter what other factors were at work. However, the strong impact of reasonable interpretations (i.e., cognitive penetration) on what is perceived suggests once again that there is a substantial difference between the laws of physics and the laws of mental processing. Apparent motion is another example of a class of experimental observations in which the response does not follow the bare bones physics and geometry of the stimulus. Instead, apparent motion appears to be exquisitely sensitive to the meaning, logic, reasonableness, and significance of the scene.

The implication of all of these reports of nonveridical perceptual experience is that the laws of physics are superceded by the laws of logical reasonableness when percepts are constructed. In fact, the contributions of the stimulus pale under the impact of the rational forces that dictate what is to be seen. Since there is no bridging concept comparable to the Cosmological Principle available to psychology, there is little hope of an axiomatic-deductive theoretical explanation of these interesting, amusing, inaccessible, and thoroughly inexplicable phenomena.

Apparent motion seems from the beginning to be a clear-cut example of a paradoxical violation of what we consider to be a necessary linear or sequential causality; perception of the cause seems to follow that of the effect. However, an alternate interpretation exists. It could be argued that although the stimuli are presented in the proper temporal order—first the center light and then flanking light—the perception of apparent motion is subsequently reconstructed after appropriate internal delays and processed in the paradoxical reverse order. Of course, this is just a hypothesis that itself cannot be tested.

3.2.2 Metacontrast

Another, and perhaps clearer, example of a paradoxical violation of sequential causality is the phenomenon known as *metacontrast*. The effect is the perceptual suppression (reduced lightness to the point of invisibility) of a preceding stimulus by a subsequent one. Metacontrast is particularly compelling, therefore, because it is a paradoxical reversal of causality in which the causal stimulus *physically*, not just *perceptually*, follows the stimulus that is affected.

Metacontrast was first described by Stigler (1910); however, it was largely ignored only to become a mainstay of perceptual research during the second half of the twentieth century. Much of the interest in it was stimulated during the heyday of simplistic neuroreductionist theories of visual phenomena (see Kahneman, 1968, and Weisstein, 1972, for detailed histories and reviews).

Figure 3.2 shows a snapshot of a typical metacontrast stimulus arrangement. The shape is not important; almost any kind of central figures flanked or concentrically circled by similar shapes can be used. The phenomenon is described as a perceptual reduction of the subjective intensity of the first central figure when the two flanking stimuli are presented shortly after the first. The suppression or masking of

Fig. 3.2 The stimulus for the metacontrast phenomena.
If the flanking forms (a) are illuminated 80 msecs before the central form is illuminated, perception of the central form (b) will be completely suppressed.

the central figure is very powerful; even well-trained and knowledgeable observers report the complete invisibility of the central figure, although it is physically identical to a similar stimulus presented without the flanking figures.

Timing is critical in determining the phenomenon; the optimum masking effect occurs at an interval of about 80 msec between the initial occurrence of the central figure and subsequent occurrence of the two flanking figures. The spacing between the metacontrasted and contrasting stimulus is less critical than the time relationships; however, the further apart are the central and flanking figures, the less the suppression of the central figure, all other influential factors being kept constant. A full discussion of the methods, stimuli types, conditions, and theories (some plausible and some highly imaginative) of the metacontrast phenomenon can be found in Uttal (1981).

There are several interesting facts about this phenomenon. One, as already mentioned, is that geometrical similarity between the contrasted and contrasting stimuli seems to be required. Another is that the contrasted (i.e., invisible) stimulus may have some cognitive effects even though it is totally imperceived by the observer. Whatever the relationships underlying metacontrast, it is a clear demonstration of a cognitive process in which the physical causal stimulus follows the physical affected stimulus. Causation, at least in a psychological sense, is operating backwards in violation of the physicist's conception of linear causality.

Another example of what appears to be a paradoxical violation of perceptual causality was reported by Eagleman and Sejnowski (2000; 2003). They were studying the line-motion illusion in which a line, all parts of which are presented simultaneously, appears to grow outward from an initially attended location. The important part of their study in the present context was that the reported illusion could be affected by stimuli that occurred after the growing line had been completely perceived and, in fact, had disappeared. Eagleman and Sejnowski proposed a cognitive explanation in which the perceptual construct is "an a posteriori reconstruction" in which the perceptual event is delayed.

However, like all previous explanations of all of these time paradoxes, this type of explanation is inadequate. All that they have really said is that the physical stimulus and the perceptual response are nonveridical, and that the explanation to account for this discrepancy lies somewhere within the vast complexity of the mind-brain system. Unfortunately, such theories are nothing more than verbal restatements of the phenomena clothed in mentalist terminology and are operationally and reductively useless. All are circularly descriptive, but not enlightening in any explanatory sense.

Regardless of what inaccessible and indefinable process accounts for these paradoxical phenomena, their importance lies in their demonstration of the disconnect between physical and psychological time—a disconnect that could not have been predicted; that is, that behavior of the perceptual system—particularly with regard to its processing of time—is not following the same rules as the physical world. Such a nonveridicality makes it impossible for us to apply the traditional deductive and inferential methods that have proven so useful in the physical sciences.

It is difficult, if not impossible, to refute any of the many plausible and ingenious explanations for these paradoxical phenomena simply because they are not really explanatory. For that matter, it is equally impossible to authenticate any one. The observational fact is that a number of our perceptual experiences appear to violate linear causality. The essential point to keep in mind is that it does not matter what the underlying mechanism is that accounts for these paradoxical phenomena. It could be a high-level cognitive effect produced by some logical inconsistency (e.g., impossible apparent motion), or it could be a simple neural effect (e.g., propagation delays of the inhibited stimulus prior to strictly neural inhibitory interactions). Whatever the underlying causes of these paradoxical phenomena, the bottom line is that the laws and properties of physical time are different from those of the mental world. This means that there is no psychological Cosmological Principle bridging the gap between behavioral "here" and the inaccessible mental "there."[3]

3.2.3 Paradoxical Visual Effects of Position

Other examples of the disconnect between laws of the physical and mental worlds can be observed in perceptual situations in which the stimulus remains constant and, nevertheless, the percept changes. We are all familiar with *motion parallax*, the apparent change in the relative depth of objects at different distances when we move past a stationary scene. This shift in position is due to the fact that objects at greater distances tend to be displaced at a lower velocity across the retina than objects that are closer. Motion parallax is a compelling cue for depth perception that stimulates the perception of depth in situations where the threshold for binocular disparity is exceeded and other pictorial cues are unavailable.

Not so well known is paradoxical or *inverse motion* parallax, a phenomenon that seems closely related, if not identical, to the "reverse-perspective illusion" (Cook, Hayashi, Amemiya, Suzuki, and Lehman, 2002). In this experiment a two-dimensional picture of a three-dimensional scene is presented to an observer. When the observer moves with regard to the picture (there is, of course, no motion parallax cue since that process requires actual differences in distance or depth), there is, instead, a perceptual experience of motion in the opposite direction to the actual motion of the observer.

The typical explanation of this phenomenon is that the observer's expectation in such a scene is so great that the observer compensates for the missing parallax by misperceiving (i.e., creating) an apparent movement in the opposite direction. As Cook and his colleagues put it in describing how the effect occurs, especially in paintings by the artist Patrick Hughes:

> We conclude that the reverse-perspective illusion is a consequence of the contradiction between the changes in visual information during observer movement and the observer's implicit knowledge concerning expected changes in the visual scene.
>
> ... Rather than deny the veracity of their own implicit knowledge, i.e., the phenomenal geometry of the visual scene (Gogel, 1990) and the expected changes in the phenomenal geometry due to ego motion, observers have a strong tendency to see motion in the artwork itself. (p. 1151)

What appears to be another manifestation of the same phenomenon is the appearance of the eyes of a two-dimensional picture following observers as they move past a portrait. This phenomenon differs from the inverse apparent motion effects in that it does not require motion; the picture's eyes always appear to be directed at the viewer even when stationary. As Koenderink, van Doorn, Kappers, and Todd (2004) point out in their extensive study of the phenomenon, all that is required is for the person depicted to have looked directly at the lens when the original photograph was taken.

In their experiment, they used a photograph of a torso (rather than a face) and showed that the phenomenon was more general then just for the apparent direction of gaze in a portrait. Their results indicated that their observers always saw the torso (or a picture) as being oriented toward them in frontoparallel view, even when looking at the picture from an oblique angle. Despite this constancy, observers were aware of the foreshortening produced by the oblique angle. Their study came to the following conclusion.

> ...the pictorial space in a painting hung on a wall and the visual space that contains the visual wall, picture frame, etc. (in many respects very similar to the physical space containing wall, picture frame, and observer) are virtually *independent*. Observers perceive the wall indeed as oblique with respect to their direction of view... [However] They always see an object depicted in a frontoparallel pose ... as facing them squarely, whatever the angle of view. (p. 526)

What this means is that the observer treats the two kinds of information (pictorial and visual—i.e., perceptual) in different ways. If something is depicted as frontoparallel (facing the observer) it continues to be interpreted as such, even though the backgrounds and surrounds are not perceived in the same way. Koenderink and his colleagues argued that this is not the result of some kind of simple compensatory "cognitive" mechanism. Rather, they suggested that the "(perceived) physical and pictorial spaces segregate" (p. 513). However separately they may be dealt with, each is responded to in a manner that is determined by the original picture, regardless of any simultaneous appearance of obliqueness or foreshortening. In other words, we see what is determined by the original *intent* of the artist or photographer and not by the true physical geometry.

Intent, like all other intangible motivational forces, is an elusive measure, of course. However, like many other illusions mentioned in this chapter, the conclusion is robust that the exact nature of the physical stimulus (i.e., the cause) is inadequate to explain the perceptual phenomena (i.e., the effect). The usual explanation of such phenomena is that the interpretation of the *meaning* of the stimulus overwhelms its physical *geometry*. Nevertheless a common logical theme emerges:

1. It is observed that a phenomenon is nonveridical with the stimulus.
2. Complex and high-level cognitive processes (e.g., reconstruction) are assumed to overwhelm the physical properties of the stimulus.
3. We cannot "explain" (i.e., derive) the phenomena from basic principles or axioms, so we simply rename them within some vague typology and attribute them to hypothetical and otherwise inaccessible processes.

This kind of explanation is not implausible, however improvable. The undeniable point being made by all of the inconsistencies between the stimulus and the perceptual responses, however, is that there is a vast conceptual, measurement, and qualitative gulf between the inaccessibility of the human mind and the inaccessibility of a far away galaxy.

As we now see, there are many other illusions that demonstrate other failures of correspondence (i.e., nonveridicality) between the properties of physical time and space, on the one hand, and perceptual experiences, on the other.

3.3 DISTORTIONS OF TIME AND SPACE

Relativistic thinking alerts us to the possibility that physical objects can be temporally and spatially distorted. For example, the gravitational field of a heavy object distorts the space around it. One of the extraordinary accomplishments of Einsteinean general relativity was the suggestion that this distortion in space-time might account for (or actually be) gravity itself.[4] As another example, his special relativity showed that the spatial dimensions of an object may appear to an observer to be lengthened when the object moves at velocities approaching the speed of light. All of these effects occur only under extreme conditions of mass and velocity in the physical domain. They are characterized by a dependency of the distortion on the temporal and spatial properties of the environment.

We also know that similar distortions of time and space occur in the mental world at the human scale without any of the extreme conditions. Unfortunately, there is no theory or promise of a theory comparable to special or general relativity to explain these perceptual nonveridicalities. However, we do know that these distortions are also dependent on how we interpret the spatial and temporal properties of the stimulus environment.

The generic name for this kind of perceptual distortion is *illusion*. Illusions are perceptual experiences that are nonveridical with the properties of the physical stimulus. Some of the best-known examples include:

- A physically straight line may appear to be curved depending on its spatial context.
- Movement may appear to reverse direction depending on prior events.
- A patch of gray may appear to be of one degree of lightness in one context, while in another, exactly the same physical stimulus may appear quite different.
- Time, physically measured in terms of a constant number of cycles of an atomic clock, may appear to be short or fast depending on cognitive activity, particularly to the extent that attention is allocated.

All of these phenomena and many other illusions or perceptual distortions depend on the relationships among the various spatial arrangements or temporal sequences of the components of the physical stimuli. In a purely lexicographic sense, they too are relative; however, the relativity is in low-level terms of energies, velocities, and separations; it would be misleading to refer to these illusions as "relative" in a way that suggests some kind of an analogy with physical relativity. A better choice of nomenclature would be *relational*. Certainly, there is no intent to suggest that they are manifestations of any kind of Einsteinean relativity theory. For that matter, there is no widely accepted general theory of any kind to account for these illusions.

These illusions are "relational" in the sense that they are all dependent on the relationships that they have with their surroundings, whether they be temporal, spatial, or even intensive. One of the most basic is the classic simultaneous contrast phenomena shown in Figure 3.3.

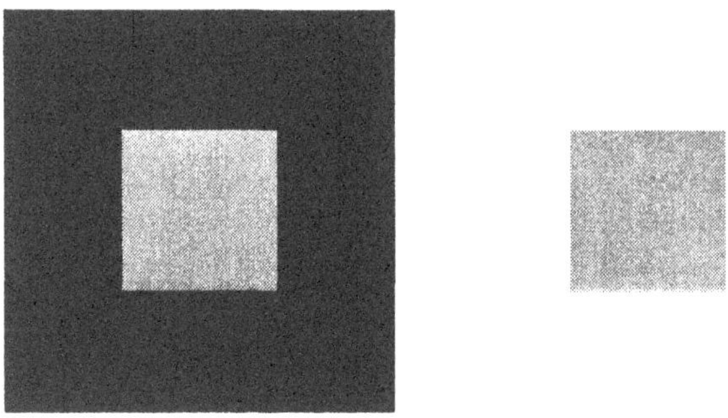

Fig. 3.3 The simultaneous contrast phenomenon.
The lightness of the central grey square depends on its surround. A darker surround will make the central square appear lighter, and vice versa.

The relational interaction exhibited in this figure between the central square region and its spatial surround is the defining characteristic of illusions. Although the two central squares are of exactly equal intensity, they are perceived as being distinctly different in lightness. The nonveridicality between the stimulus and the response (due, in this case, to spatial relations) is the defining characteristic of all illusions, whether they are distortions of spatial form, lightness, color, or movement.[5]

Among the most important theoretical points made by all illusions—visual, auditory, and even tactual—are:

- The perceptual responses are not congruent or veridical with the physical stimulus.
- Illusions are influenced by: (a) the meaning or semantic content of the stimulus; or (b) by their relations with other parts of the stimulus environment.
- Whereas the physical stimulus may be precisely defined by quantitative measurements, the perceptual world seems to operate by rules and logics that are different from those of the physical world. In general, they are not predictable a priori or susceptible to direct quantification, and, therefore, all measurements of the mental properties are suspect.
- In general, therefore, visual illusions have not and probably cannot be predicted or explained by physical principles and laws alone.
- It is, therefore, not possible to predict how a particular set of stimuli is going to be manifested as a perceptual experience.

A corollary of this argument is that just as we have no way of predicting how a particular stimulus arrangement plays out perceptually, we have no way of authenticating any of the many possible theoretical explanations of an observed illusion.

At the risk of overkill, but in anticipation of an oft-encountered criticism of this point of view, I must reiterate an important caveat. Nothing I say here about the differences between the physical world and the cognitive or phenomenal one is intended to imply anything supernatural. Nor am I suggesting that the mental processes are anything less than the outcome of the activity of a material system—the brain. The only point being made here is that these *mental processes* operate by rules and laws that are not immediately derivable from those of the physical world. The main reason for this situation lies not in any questionable difference between the reality of the two domains, but because of the extreme complexity of the mental world.

Thus, it would not have been anticipatable on the basis of any form of physical or mathematical derivation, for example, that the Muller-Lyer illusion (See Figure 3.4) should occur. This illusion (and all others) is a result of the extreme complexity of the mind-brain system's functioning. The rules that the mind-brain system follows are not implicit in either Newtonian or Einsteinean models of physical re-

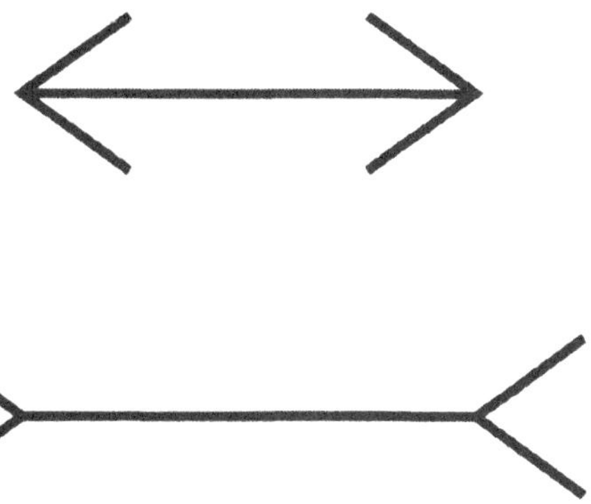

Fig. 3.4 The stimulus for the Muller-Lyer Illusion.
The two lines appear to be different lengths depending on the orientation of the end lines.

ality. It is a sobering thought that not only is there no equivalent model of visual illusions currently existing, but there is no likelihood that such a comparable axiomatic-deductive theoretical structure could ever be developed. We have no basic axioms comparable to F=ma and no system of mental processing rules comparable to *dy/dx* that would permit such predictions. All we can do is observe and describe.

The discrepancies, nonveridicalities, or illusions mentioned here are among the best known of all psychological phenomena. Many fanciful theories have been generated to explain various illusions at both the cognitive and neural levels. The important point is that they are not predictable from the usual laws of physics, and, therefore, are both phenomenologically surprising[6] and theoretically impenetrable; in other words, inaccessible even to some Einstein-like superpsychologist who might appear in the future.

The inaccessibility barrier created by the absence of a psychological Cosmological Principle cannot be overcome just by collecting additional data about phenomena such as illusions. Instead, it a basic, fundamental, and in principle barrier that may never be ameliorated by any empirical or logical machinery. The bottom line is that although there is no question that mental processes are the result of physical (i.e., neurophysiological) activity, conventional explanatory approaches do not permit us either to predict the phenomenal output from the stimuli or to infer underlying mechanisms from the responses.

3.3.1 Nonveridicalities of Spatial Perception

Static geometrical or optical illusions come in an enormous variety of different types and kinds. All of them appear to distort the geometry of a part of the stimulus scene because of the proximity of, and relationships with, other parts. Many occur without requiring any movement or special shading and are typified by the misreading of curvature, angle, size, or some other aspect of the environment. An idea of the variety of these forms can be gleaned from the three plates of Figure 3.5.

Because of the variety of these illusions, it has proven extremely difficult to develop some kind of an organized classification system or taxonomy, much less a comprehensive theory, for them; each illusion is seemingly idiosyncratic. The usual strategy has been to collect together those that involve, for example, distortions of angles, size, length, or direction. A further problem is that some of the illusions that are perceptually similar seem to be based on different errors of interpretation. For example, the Baldwin and Muller-Lyer illusions both produce perceptual elongations of one line with respect to another, but the defining properties (the size and direction of the wings) are different and, therefore, some of the offered explanations are unrelated.

A major, and almost unique, exception to this dearth of taxonomic classification of illusions was the work of Coren, Girgus, Erlichman, and Hakstian (1976). They carried out an extensive series of experiments, the findings of which were factor-analyzed into what appeared to be five distinct categories. Although the utility of their classification system is to be admired, it must be remembered that it is a taxonomy of phenomena and not of explanations or theories. Thus, the possibility remains that what appear to be alike perceptually may be due to quite different causes.

1. Line direction and shape distortions
2. Size contrast
3. Illusions of overestimation
4. Illusions of underestimation
5. "Frame of reference illusions"

Because the actual mechanisms and processes that account for these illusions are buried in our inaccessible mental processes, I argue that it is impossible to develop a comprehensive theoretical explanation of any of them. What we have instead is a "grab bag" of individual and isolated microtheories. Many of the proposed "theories" of a particular illusion are not even superficially explanatory. Instead, many are obviously little more than circular restatements of the phenomenology (e.g., "We see the illusion because we misperceive the angles"; Robinson, 1972). Some suggest that illusions are the result of cognitive "interpretations" of the meaning of the stimulus configuration, an approach that is, once again, tantamount to just describing the phenomenon. Of two things only can we be sure: (1) Illusions are the result of our brain's processing of physical stimuli; and (2) They are misinterpretations or distortions of those stimuli.

Other putative theories of illusions seek to invoke (or invent) special "cognitive" laws and rules to describe hypothetical (but otherwise invisible and inaccessible) processes that might account for the illusions. For example, the lines in the Poggendorf illusion are often described as representing convex and concave cor-

Name	Illusory stimulus	Effect
Vertical – horizontal	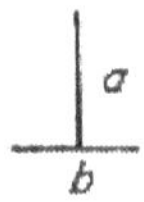	*a* appears longer than *b*
Müller – Lyer	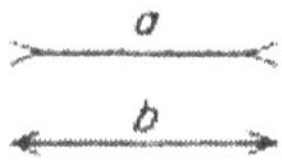	*a* appears longer than *b*
Fat lady	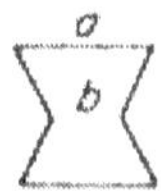	*b* appears greater than ½ of *a*
Convergence – divergence	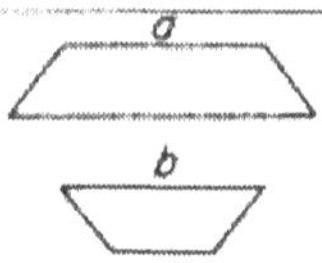	*a* looks longer than *b*
Wundt's area	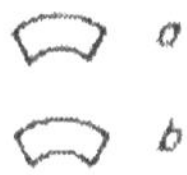	*b* looks larger than *a*
Filled – empty space a	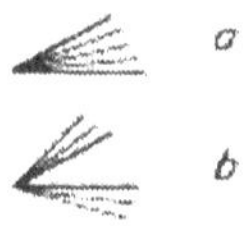	*b* looks larger than *a*
b		*b* looks longer than *a*
Curvature		Even though all arcs have same curvature, shorter appears less curved
Poggendorff	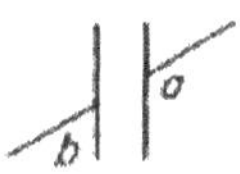	*a* nd *b* do not appear to be colinear

Fig. 3.5a Stimuli producing spatial illustions (Uttal, 1981).

Name	Illusory stimulus	Effect
Zöllner	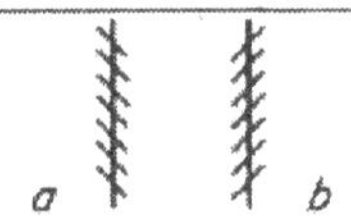	Lines a and b appear nonparallel
Ebbinghaus (size contrast)*	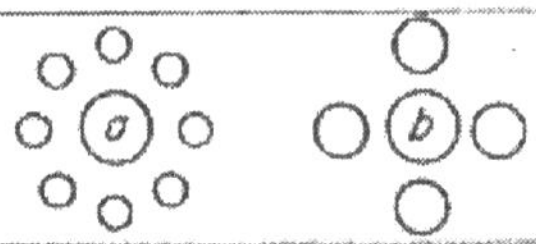	a looks larger than b
Angle contrast		∠a looks larger than ∠b
Perspective	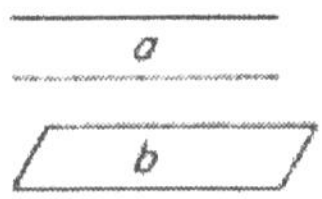	Lines in a look further apart than lines in b
Wundt's lines†	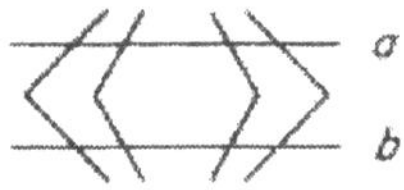	a and b look curved
Wundt's curves		Line a looks curved
Winged lines		a looks longer than b
Ponzo	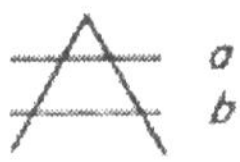	a looks longer than b
Divided line	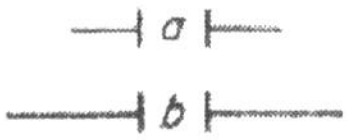	Gap a looks larger than gap b
Baldwin	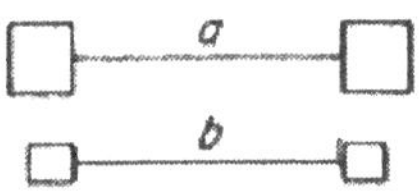	b looks longer than a

Fig. 3.5b Stimuli producing spatial illustions (continued).

Name	Illusory stimulus	Effect
Rod and frame	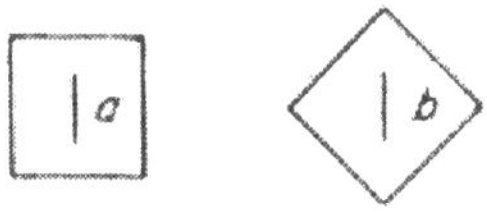	*b* looks nonvertical
Schumann square	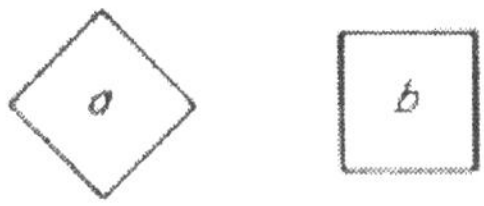	*a* looks larger than *b*
Shepard	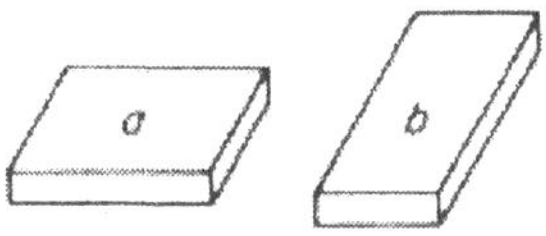	*b* looks larger than *a*
Jastrow–Lipps	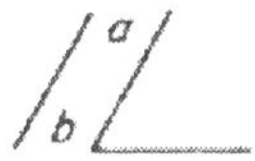	Lines appear closer at *b* than at *a*
Sander parallelogram	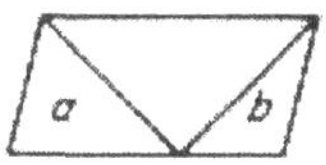	*a* looks longer than *b*
Orbison	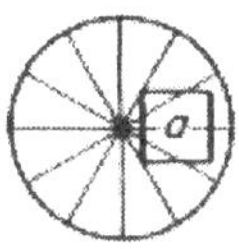	Square *a* distorted in parallelogram

Fig. 3.5c Stimuli producing spatial illustions (continued).

ners of otherwise incomplete three-dimensional structures (Zanuttini, 1976). The perceptual distortions in this case are assumed to arise from the fact that the two-dimensional projections of three-dimensional objects are indeterminate and permit alternative (and often incorrect) interpretations. Therefore, our "minds" fill in the missing information or introduce constraints that reduce the problem from an ill-posed one to at least one plausible solution of many possible ones. Unfortunately, the introduction of this additional information permits our perceptual system to create or construct distorted interpretations.

Other theories invoke fields of mental "forces" that have one part of the visual stimulus exerting distorting forces on other parts, analogous to the operation of gravitational or electromagnetic fields. So-called "spatial contour" models, for example, were proposed by Fisher (1973). Similarly, neuroelectric fields were a mainstay of the Gestalt tradition and are reflected in the many modern field theories of mental activity so popular today. Theories of "force fields" are often replaced by attributing these illusions to the activity of single neurons whose individual behavior often seems to mimic the global perceptual process.

All such neuroreductionist theories of illusions, however, are feeble attempts to leap from the wonderful accomplishments of modern day neurophysiology to the inaccessible and mysterious domain of perceptual illusions. Although they were extremely popular a decade or two ago, common sense seems to have exerted an intellectual "force," and such theories are less likely to be found in the present perceptual literature. Again, there are exceptions, such as the view expressed by Spillmann (1999):

> delighting in pure phenomenology without considering known physiological findings for constraining models of visual perception does not get us any closer toward understanding the underlying mechanisms. (p.1491)

However, it appears that simplistic neuroreductionist explanations of perceptual phenomena are rarer nowadays than in previous decades. Although neurophysiology may constrain some "models of visual perception," by no means do they provide enough information to constrain explanations of unique theories or models. The implications of the emerging consensus that we are not going to move theoretically from neural data to perceptual theories should now be clear.

An alterative to Spillmann's assertion, which may not appeal to those who seek neural explanations of perceptual phenomena, may be proposed:

> Since no known physiological mechanisms can adequately constrain models of visual perception, this is not a route toward understanding the underlying mechanisms. (Uttal, de novo)

Others seek "explanations" by interpreting one illusion in terms of another. Thus, for example, Day (1972) and Gregory (1963) suggested that many illusions in which linearity is distorted may be due to the effect of other illusions such as size constancy. Others are much more mathematically sophisticated and use non-Euclidean ideas from relativistic physics along with some supplementary assumptions about psychological space to develop mathematical models that describe a selected range of these illusions. The works of Hoffman (1966) and Watson (1978) are especially notable in this regard.

A novel approach has been proposed by Purves and his colleagues (e.g., Purves, Lotto, William, Nundy, and Yang, 2001; Howe Yang and Purves, 2005; Howe and Purves, 2005a; Howe and Purves, 2005b). The suggestion made by this group was that the statistics of the world around us provide a probability distribution of possible responses, and that previous experience leads us to choose from among these possibilities the most probable. The final perception, therefore, they argued, is a combination of the statistical probabilities of the environment and the choice behavior of the observer.

Purves and his group also emphasized that the nature of illusions "cannot be deduced from principles of projective geometry per se" (Howe, Yang, and Purves, 2005, p. 7711). In other words, they agreed that the normal rules and laws of physical space and geometry do not apply to psychological space. Their statistical approach also suggests that a purely neuroreductive explanation is not likely.

Although it is certainly true that illusions are nothing more or less than a pattern of neural activity, there is no theoretical bridge with which we can yet cross from that neural activity to the statistical properties of visual space. Ultimately, all we have recourse to is the reported phenomenology of the illusions. Nor is there any way to bridge the gap between the phenomenology and the cognitive processes that presumably underlay all of these illusions. This is another aspect of what is meant by cognitive inaccessibility.[7]

There are many other illusions in which the shapes of objects are distorted or their position in space misinterpreted. Although it is impossible to list them all here, I cannot resist mentioning one in particular—the illusion of the rising pitched fastball in baseball. It is commonly reported by baseball players that some high-speed pitches appear to jump up at the end of their trajectory, creating an especially difficult hitting task for a batter.

McBeath (1990) studied this phenomenon and makes some highly important points about the rising fastball. He notes that whereas curved, sliding, and even irregular trajectories are physically possible because of uneven air flow induced by spinning, no comparable physical explanation can account for the rising fastball. This pitch appears perceptually to hop up just before it crosses the plate by as much as a third of a meter. Despite the fact that no physical evidence of such a trajectory has ever been recorded or even suggested, it is a deeply held belief on the part of batters.

McBeath concluded that the "hop" just before the ball reaches the batter is totally illusory. It is produced by an underestimation of the position of the ball by the batter early in its trajectory; during that early phase the ball appears to be lower than it actually is. In the last few meters, however, the actual speed of the fastball becomes perceptually appreciated, and the batter interprets this reevaluation of speed as a rise in its position. There are no violations of Newton's laws, but there are definite contradictions between the trajectory of the ball and its perception.

This is another classic example of cognitive penetration—the effect that our interpretations, judgments, and biases can have on our perception of events. These "penetrations" are so powerful that they can actually override the predictions made by the laws of physics! To put it in a catchy little phrase: the laws that work "there" do not work "here."

3.3.2 Nonveridicalities of Temporal Perception

The argument made in the previous section is that unpredictable, unanalyzable, and distorted phenomena occur in our perceptual experience of space. These distortions are not predictable because they seem to invoke cognitive processing mechanisms that operate by rules and laws that are different from those of the physical world. In fact, I suggest, there is not currently, and there may not be in the future, any comparable set of rules and laws that can pertain to the mental domain. Illusions are unanalyzable for many reasons, but among the most important are the sheer complexity of the mind-brain and the underdetermined nature of our behavior vis à vis the underlying mechanisms. Therefore, it is not possible to infer from the reports of the distortions what causes them.

The examination of visual illusions in the previous section emphasized the distortions of spatial patterns as a result of simultaneous spatial relationships of the objects in the visual scene. It also raised the problem of how our cognitive beliefs, expectations, and reasonableness can affect our perceptual experiences. We call this process *cognitive penetration*.

Motion Aftereffects: I now turn from consideration of what were mainly spatial distortions to some that are primarily temporal to make the same general point: The physical implications of the stimuli do not always result in a veridical perceptual response. It is equally well established that temporal as well as spatial relationships can also play a complex role in determining what we see and how we subjectively evaluate the passage of time. Temporal illusions have been well known for millennia. Indeed, Aristotle refers to one of the most familiar of motion aftereffects:

> when persons turn away from looking at objects in motion, e.g., rivers and those that flow very rapidly, they find the visual stimulations still present themselves for things really at rest are then seen as moving. (Cited in McKeon, 1941, p.621)[8]

This section deals with illusions of this genre and related unpredictable and inexplicable phenomena in which time and motion are distorted. The basic observation is that preceding temporal events play a role in determining what we perceive. As I shortly conclude, however, there are—like the spatial illusions of the previous section—an enormous variety of unconvincing theoretical explanations of temporal illusions. Disappointingly, there is neither consensus nor convergence on consensus that any of the explanations so far proposed are going to be fruitful in understanding these temporal illusions.

To begin, let's deal with a set of visual illusions in which spurious movements or shape distortions are induced by preceding "conditioning" events. The interrelationships between these antecedent events and the resulting illusions are characterized by the persistence of the effect of the conditioning over a prolonged period of time, but with a progressive decay in efficacy. These illusory effects of conditioning produce effects opposite to the direction of the conditioning stimulus.

The prototypical reversed motion illusion (now known as the waterfall illusion) discussed by Aristotle is the most famous example of this type of illusion. After viewing a waterfall, if one turns one's attention to an adjacent nonmoving wall of solid rock, the rock itself appears to be flowing upwards. Other well-known effects of conditioning include the distorting effects of previewing particular geometrical forms. For example, after a conditioning period in which a line curved to the left is inspected, a physically straight line appears to curve to the right. Figure 3.6 presents a group of these time-based illusions.

Unlike the spatial illusions depicted in Figure 3.5, these figural aftereffects are much more like each other in that they seem to share a common property: an inverse motion response to a previously viewed moving stimulus. As a result, there has arisen a very popular, common, but extremely difficult to test, theoretical explanation for all of them: neuronal fatigue. The usual expression of this theory is that our perception is determined by a balance between activities in two opposing neural subsystems. These mechanisms purport to balance our perception of motion to the left and right, up and down, etc. They are supposed to be comparable to the well-authenticated opponent color neurons observed by such pioneers as DeValois, Smith, Kitai, and Karoly (1958) in the lateral geniculate body, and by MacNichol and Svaetician (1958) in retinal bipolar cells. Since there was suggestive, but indeterminate, psychophysical evidence that color perception was also mediated by opponent mechanisms at least someplace in the visual system, the idea of neural opponency, of a balance between competing mechanisms, gained wide currency in the last half of the twentieth century.[9]

It was a small step, therefore, to propose that our spatial, as well as our color vision systems, operated on the basis of a balance between opponent mechanisms. Thus, for example, we are supposed to see straight lines because of a balance between left-curving and right-curving neural mechanisms. Should we overuse, fatigue, or exhaust one or the other of these two opponent mechanisms (for example, by prolonged viewing of a left-curving line) we would subsequently spuriously perceive a straight line as a right-curving one.

Other theories have come and gone; however, the neural fatigue theory became especially prominent in the heady days of neurophysiological explorations and continues to populate semi-popular discussions about why we see these illusions. Unfortunately, these superficially simple phenomena are not as simple as they may at first appear. A further hindrance to blithely accepting neural fatigue theories is that the opponent neural mechanisms for geometry, so frequently invoked, have never been definitively identified by electrophysiologists.[10]

The main argument, however, against the hypothesis that figural aftereffects are simply a result of fatigued neurons can be found in one of the most famous of all such phenomena: the contingent negative aftereffect discovered by Celeste McCollough (1965). McCollough carried out an experiment in which complementary color aftereffects were determined not only by the wavelength of the conditioning stimulus, but also by the direction of the lines in it. Thus, the final perceptual response was said to be "contingent" on the interaction between these two stimulus variables.

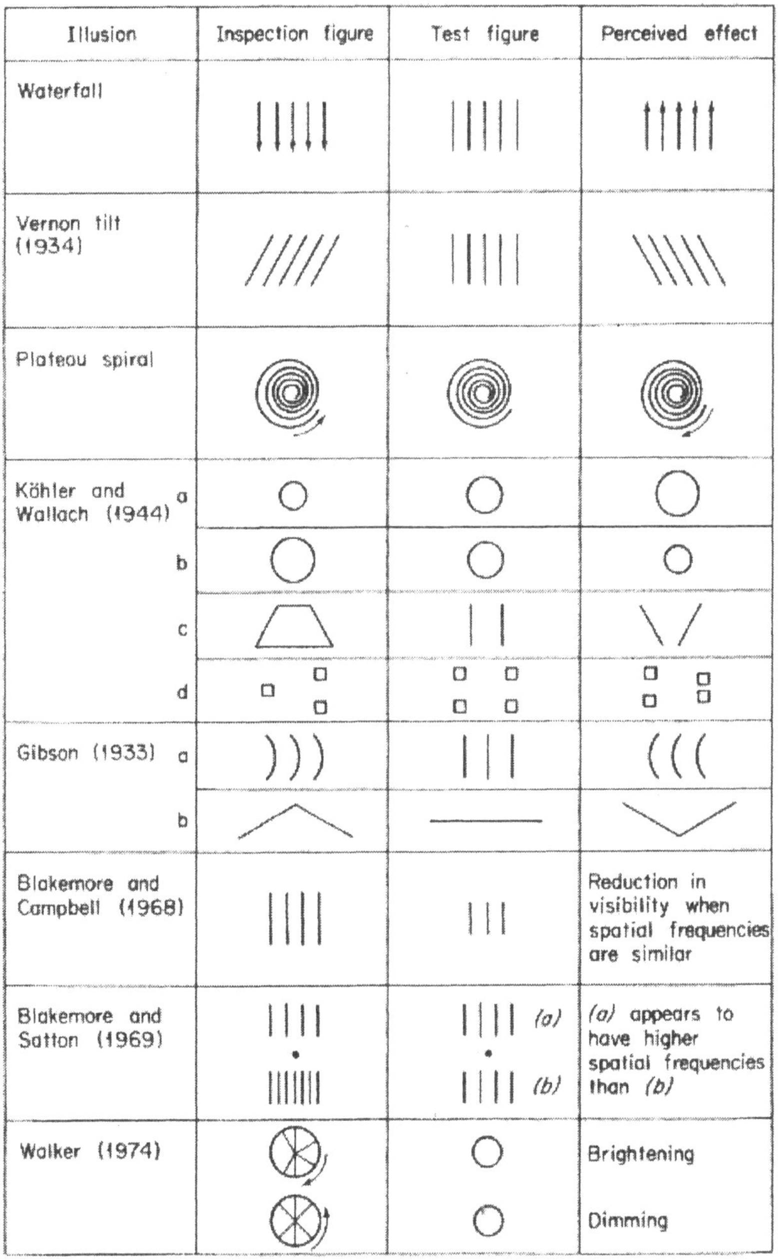

Fig 3.6 Stimulus producing temporal illustions (Uttal, 1981)

The McCollough effect has been enormously influential in visual science. A very substantial body of empirical knowledge has accumulated concerning its phenomenology and the influence of a variety of stimulus conditions on it and that of related contingent phenomena. An excellent review of the literature can be found in Humphrey and Goodale (1998).

Although McCollough's original explanation was framed in terms of the fatigue of neurons that were sensitive to both orientation and color, especially those in the primary visual regions, this explanation has encountered serious challenges since the time of her pioneering experiment. One notable challenge is that fMRI studies have been carried out that suggest that widely dispersed regions of the brain are involved when the phenomenon is elicited (Barnes, Howard, Senior, Brammer, Bullmore, Simmons, and David, 1999). Although the fMRI may have severe limitations in identifying specific locales for particular psychological processes, in this case the absence of any regions that can be uniquely associated with families of "fatigued neurons" argues against the idea of any kind of simple opponent mechanism accounting for the phenomenon.

A far more serious challenge to any neural fatigue explanation of the McCollough effect is that the aftereffects of the contingency persist over many months if not tested in the intervening period (Jones and Holding, 1975; Riggs, White, and Eimas, 1974). The contingent nature of the basic phenomenon and its long-term persistence strongly suggest that it is not a simple result of fatigued neurons. Simple fatigue certainly would have recovered during the prolonged periods that the effect can persist.

Theoretical discussion of the McCollough effect in particular, as well as other visual aftereffects in general, have invoked such widely diverse phenomena as the activity of neurons, cognitive interpretations, and various forms of learning and memory. Clearly, no one knows what is the effective cause or source of this phenomenon. Like many other mental processes, there are simply not enough empirical anchors, too many barriers to accessibility, and a gross underdetermination of underlying mechanisms by the observed phenomena to permit the kind of axiomatic deductive analysis so successful in physics. What is clear is that aftereffects of conditioning stimuli are additional examples of unanticipated (if not unpredictable) events. That is we would not have anticipated that any of these illusions would occur from what we know about the stimuli. The laws of either neurophysiology or the external physical world are of no help. Thus, temporal aftereffects cannot be simply attributed to exhausted neurons; the interrelationship or symbolic relationship of the stimuli certainly also determines the nature of the aftereffect.

This counterargument raises the possibility that other noncontingent figural and temporal aftereffects represent the outcomes of much more complex and equally inaccessible cognitive and neural processes and mechanisms than simple neural fatigue.[11]

As it stands now, there still is no comprehensive theory, neural or high-level cognitive, that even begins to satisfactorily explain the extreme nonveridicality observed in these illusory aftereffects. No matter how much the fatigued neuron theory is stretched, it remains inadequate. We are probably just as far from explaining these highly amusing and interesting phenomena as we were a century or more ago when they were first observed.

Subjective Time: Our perception of time itself can be distorted in ways that suggest that there is also a vast gulf between the laws of cognitive processes and those of physics. For example, our perception of time is extremely elastic. How long an interval is depends on a host of personal and environmental conditions, including something as elusive as allocation of our "attention."

Physical time, on the other hand, tends to be continuous and homogeneous. A second now is a second later for observers within their own frame of reference. In psychology, however, a "moment" or an "instant" can be extremely variable. Furthermore, although we generally preserve ordinality in time, there are many instances in which even order can be violated. (See the discussion on temporal paradoxes earlier in this chapter.) Backward looking memory and forward looking anticipation can also, from some points of view, violate the ordinality of time. Although perceptual time is roughly continuous, this is not a universal attribute of all humans or of all human activities. Discontinuities in time due to sleep or highly directed attention are common examples.

As an even more extreme example of the variability of subjective time, the Pirahâ people of the Brazilian rain forest seem to have little concept of the past or the future. Everett (2005) asserts that:

> Their grammar and other ways of living are restricted to concrete immediate experience (where an experience is immediate in Pirahâ if it has been seen or recounted as seen by a person alive at the time of telling), and immediacy of experience is reflected in the immediacy of information encoding—one event per utterance. (p.632)

Everett goes on to note that these interesting people have very few words for time. Their complete vocabulary for time consists of ten event-related terms, such as "ahoakohoaihio," which means early morning and literally is translated "at fire inside eat go." There are no words for tomorrow or yesterday beyond "ahoapio," which means another day and literally "other at fire." At the other extreme are westerners, especially those who live in northern hemisphere cultures who depend on precision clocks and for whom being "on time" is a highly regarded virtue.

For some, as I already noted, a kind of time machine is embedded in our memories. We can recall events from times past in a way that brings them to the present. The classic work of Penfield (1955) highlighted the ability of the brain to relive (albeit in a dream-like flashback) past experiences when it was stimulated by electrical currents.

To carry this metaphor further, ambition, like memory, is also a time machine. People anticipate future rewards and shape their behavior appropriately. Although a little far-fetched, anticipation can be considered to be a violation of linear causality; events (at least, conceivable or hoped-for events) in the future influence current behavior. Some psychologists have pointed to this planning or anticipating as a fundamental property of humanity. However, as usual there appear to be exceptions. I have already alluded to the Pirahâ people as having little sense of the future. Nevertheless, it is likely that they do some planning; at a minimum, it seems that it would be totally maladaptive and possibly lethal not to carry a tool or weapon when one goes out on a hunt, or not to store some food for the next day.

The study of such anticipatory behavior has recently been rejuvenated, especially in the discovery that it is not entirely restricted to humans. Mulcahy and Call (2006) for example, showed that some of the great apes behave in the same way: bonobos and orangutans saved simple tools for future use. Although the major contribution of their work was in the theory of the evolution of behavior, it also illustrates that these great apes, albeit to a lesser degree, share with humans the ability to transcend the limits of physical time.

Unfortunately, given the enormous amount of interest in it, both subjective and physical time research are clouded by the difficulties in specifying exactly what it is that we mean by physical "time" (see page 4), as well as the significance of the results of an altered and flexible perception of time. Philosophers, psychologists, and physicists have all debated the nature and meaning of time for centuries. David Hume and Henri Bergson believed in discrete perceptual units. William James thought that psychological time flowed continuously, in a "stream of consciousness."[12]

Clearly, however, whatever time is and whatever it means, there are some empirical facts that suggest that people distort time (or their perception of it) in profound ways. The following paragraphs give an idea of how physical time is not always congruent with perceptual time.

Simultaneity and Temporal Order: One of the first clues that physical time and psychological time are not congruent can be found in the findings of experiments that deal with apparent simultaneity. Two stimuli that are close together in time are perceived as being simultaneous. This lack of temporal resolution is evidenced in a number of psychophysical phenomena, including masking (Uttal, 1969), form recognition (Eriksen and Collins, 1965), and temporal order judgments (Hamlin, 1895; Smith, 1933).

Of these, temporal order judgments are among the most interesting because the order in which stimuli are perceived appears to vary depending on the properties of the stimuli. In Hamlin's and Smith's classic studies of the temporal order of simultaneously presented acoustic and visual stimuli, it was discovered that either stimulus could perceptually precede the other depending on their relative intensities. (Recall also the paradoxical results of apparent motion and metacontrast discussed earlier)

Thus, for reasons that must in some ontological sense ultimately be accounted for by the neurophysiological properties of the central nervous system, the actual explanations of how we order events in time are likely to remain mysterious. A violation of order is, perhaps, the most extreme violation of the properties of physical time. One can appreciate, if not explain, elastic (but topologically consistent) distortions of time as a function of attentive effort; however, the perceptual violation of the order of stimulus events is a phenomenon of much greater import.

There is no question that such an extraordinary phenomenon as a reordering of events can occur and is very difficult to explain. Neumann and Niepel (2004), for example, recently compared reaction times and temporal order judgments and concluded that although the reaction time is probably determined by physiological latency times, temporal order effects seem not to be explicable in the same manner. They concluded:

> Our survey of the available data has suggested a different picture. The TOJs [Temporal Order Judgments] are clearly not generally based on the latency difference that the RT data predict. The majority of the experiments even show a reversal of the predicted modality-based difference, and the few discrepant findings can at least in part be explained as hidden intensity effects. (p. 255)

This kind of violation of physical temporal order by the perceiver and the incompatibility of the results of such experiments with the most obvious explanation (differential latencies) is a nonveridicality of exceptional significance. Although Neumann and Niepel obscure the possible significance of this set of findings by their use of the word "hidden," what is really implied by their results is that the rules of temporal order perception are quite different from those used to describe the properties of the physical stimuli. This is another example of the distortion, discontinuity, and even perceptual reversal of time as it is measured in the physical world.[13]

Elastic Time: Although the reversal of temporal order judgments just discussed are striking examples of how physical time and perceptual time can become disconnected, there are many other phenomena in which perceptual time appears to be elastic. That is, although order may be maintained, a homogenous sequence of equal intervals (a property of physical time) has often been shown to be inaccurate as a descriptor of psychological time. A useful summary of the kinds of situation that can lead to either an extension or a contraction of perceived time has been offered by Friedman (1990):

1. Absorbing tasks shorten the impression of time in passing.
2. A greater number of events lengthen impressions of a given duration.
3. An interval seems longer if one knows in advance that it is to be judged.
4. We experience an acceleration of the passage of time as we grow older.
5. An interval of time seems exaggerated if we are frustrated with waiting, anticipating a pleasant experience, perceiving ourselves to be in danger, or carefully watching for some event to occur.

6. An interval seems longer if we remember more of its contents or if it was made up of more distinct segments. It seems shorter if we think of it in a simpler way. (p. 20)

Friedman goes on to note that there is solid research that "subjects give longer estimates in prospective than in retrospective conditions, even when the duration and contents are identical" (p. 21). Friedman attributes these elastic distortions and unequal intervals of time to how people allocate their attention. That is, the common feature of all of the phenomena of time dilation or contraction considered here is how much effortful attention we attach to the intervening tasks.

Again, this cognitive explanation actually does not help to explain the neural or information processing functions that account for the elasticity of perceived time. What it does do, along with many of the other topics on time perception in this chapter, is attest to the fact that psychological time does not exhibit the same properties as physical time. Psychological time is neither homogeneous, isotropic, nor monotonic and often discontinuous! Indeed, we may extrapolate from these discrepancies to suggest that many of these nonveridical results argue that the properties of time perception necessary for quantification are not, in general, present. Without quantification, measurement becomes elusive, and without measurement, scientific explanatory theories become, at least problematic. Just how "problematic" is clearly illustrated by the paucity of solid theory in this field.

A further elastic distortion of time occurs in highly stressful situations (e.g., automobile accidents) according to popular mythology. Unfortunately, this would be an extremely difficult situation to study in a controlled manner for ethical and technical reasons. Attempts to use other phenomena, such as eye movements or responses to simulated moving stimuli, cannot adequately simulate the traumatic time-distorting effect of an accident. Thus, the popular notion of the slowing of time that is supposed to occur in these stressful situations remains poorly understood.

On a more microscopic level, estimates of short durations also show how poor we are at clocking time and how complex are the laws of perceived time. A classic study (Woodrow, 1934) presented a sound or a light and asked observers to reproduce the duration of that stimulus. For short durations (< .6 sec), there was a strong tendency to reproduce a duration that is longer than the stimulus; that is, to overestimate the duration. For longer durations (> .6 sec), there was a tendency to reproduce a duration that is shorter than the stimulus; that is, to underestimate the stimulus. For stimuli around .6 sec, people seem to be much more accurate and veridical.

The point of this kind of experiment is that a simple answer based on some ineffable and indefinable property such as attention is totally inadequate to explain the subjective elasticity of time. Indeed, in Fraisse's (1963) words, "there are different laws of perception for each of the three categories of time" (p. 118).[14] This is hardly suggestive of a world in which the laws of physics are congruent with those of behavior.

3.4 DISCONTINUITIES IN TIME AND SPACE

So far we have seen how our time perception exhibits substantial elasticity and, in a surprising number of cases, even violates the sequentiality ordinality of time. There are, however, other examples of nonveridical perceptual responses that also speak to the differences between physical and mental temporal laws. I group these phenomena under the rubric of perceptual discontinuities. All of these results are presented to support the argument that the main explanatory problem faced by cognitive or mentalist psychologies is their inability to assume that the laws that work "here" also work "there."

We now are fairly well convinced that physical time is continuous, and that it does not exhibit discontinuities at the human scale. To the degree that mental time is not continuous, there is a glaring clash of the laws and properties that govern the two domains. This section examines two examples of discontinuity in our perceptual experience: the classic reversible figures and the newly discovered phenomenon of change blindness. The first is an example of how discontinuities can occur even when the stimulus is constant in time. The second is an example of how discontinuities in the stimulus world can go completely unrecognized by the human observer. Both reflect surprising anomalies of human perception, emphasizing the fact that mental time is not always congruent with physical time.

3.4.1 Reversible Figures

Reversible figures are probably next in popular familiarity to the geometrical illusions discussed in Section 3.3. They appear in every introductory psychology text as examples of the mysteries of human perception and are well known to the lay public. Two of the most famous, the Necker (1832) cube and the Rubin (1921) vase are shown in Figure 3.7.

Unfortunately, as familiar and popular as these reversible figures are, their cognitive bases are totally unknown. What we do know about these figures is that they are typically incomplete or ambiguous (underdetermined) and, thus, are amenable to alternative perceptual reconstructions depending on what interpretations and constraints are applied to them by the observer's cognitive system.

The Necker cube, for example, is a two-dimensional projection of a three-dimensional cube. In being so projected, information is lost about the three-dimensional object's actual orientation; it could be the projection of either a left- or a right-facing cube. When viewed, there is a powerful tendency on the part of the observer to apply unknown and perhaps unknowable, mental constraints that allow either of these two interpretations to be perceived. However, these perceptual alternatives can never be simultaneous, and the result of the ambiguity is a discontinuous and abrupt switching back and forth between the two alternative reconstructions. Although the particular interpretation (left or right) can be biased by directing one's attention to one or another of the corners (Kawabata,

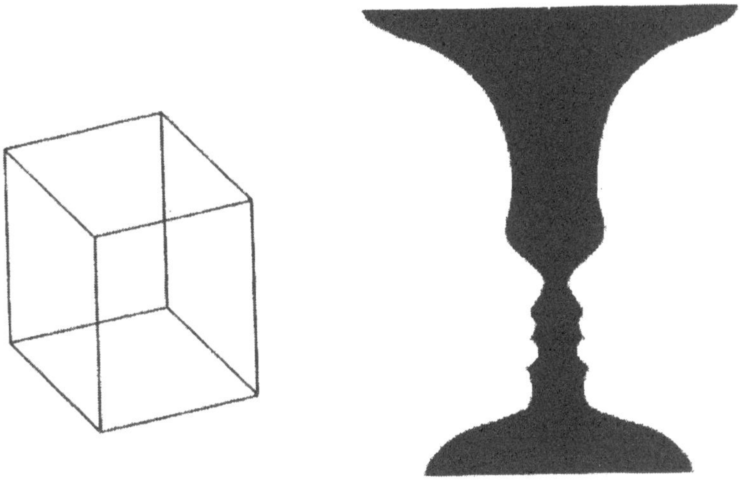

Fig 3.7 The stimuli for the Necker (a) and Rubin (b) reversible illusions, respectively.

Yamagami, and Nokai, 1978), the result is a series of "catastrophic" reversals of the two possible interpretations. In other words, there is a discontinuity in the time course of the perceived experience that runs counter to the actual physical stability of the stimulus. At no point are the two reconstructions simultaneously present, nor is there any intermediate state. An observer reports going abruptly from a left-facing cube to a right-facing cube. The flow of perceptual time, unlike that of physical time, is discontinuous.

3.4.2 Change Blindness

If the sudden reversals of perceptual experiences resulting in discontinuities in time can occur even though the stimuli remain constant, what would the reverse situation be? The best answer to this question is to note situations in which substantial discontinuities in the stimulus can occur without the observer being aware of them. Some of these invisible discontinuities are well known in traditional psychophysics. Some are novel recent results that much more dramatically demonstrate that the human perceptual system is extremely insensitive to interruptions in the smooth flow of physical time by our inability to process these discontinuities.

At the microscopic end of the time scale, we are not able to resolve two stimuli as being separate events unless they are separated by a threshold amount of time. This limit on temporal resolution (which may be of the order of 10 or 100 msec depending on the experimental situation) means that time seems to flow continuously from moment to moment, in spite of the fact that there may be actual physical discontinuities in the stimulus sequence. The most common manifestation of this effect is that which occurs when one is watching a movie or a television screen. Although a close examination of the sequence of events clearly indicates that a "moving picture" is really a sequence of still images, our perceptual system adds what appears to us to be continuity and smooth motion. Indeed, the power of our perceptual system is so great that even the physically blurred image in each frame is sharpened to perceptual clarity.

At an intermediate time scale, the modern argument has become crystallized in the debate between those who believe in discrete psychological moments and those who argue that a perceptual "window" continuously slides along the time line. Thus, scholars such as Stroud (1949) suggested that our perceptual experiences were compartmentalized into constant "psychological moments", chunks, or units (comparable to a line of separate box cars in a train) lasting for about 100 msec within which any occurring events would appear to be simultaneous. Others, most notably Shallice (1964), suggested that time is actually arranged such that there is a sliding window of variable efficacy[15] continuously moving along the temporal dimension, unlike Stroud's constant efficacy "box cars." According to Shallice, events appear to be simultaneous with varying probabilities depending on how close they are to each other within these sliding windows of time.

Unfortunately, like so many other perceptual phenomena, this debate is carried on in academic journals and circles without any sign of resolution. With the recent upsurge in neuroscientific technology, some rather fanciful ideas concerning neural oscillations have been proposed that some authors (e.g., VanRullen and Koch, 2003) suggest might speak to the problem; however, even these supporters of such neurophysiological hypotheses acknowledge that:

> It seems surprising that such a fundamental question as whether conscious perception occurs in discrete batches or continuously has not been definitively answered one way or the other. (p. 212)

Unfortunately, the expectations that this problem can be resolved by EEGs, multiple microelectrodes, or fMRI images seem wistful hopes rather than promising lines of research. Neurophysiological hypotheses of this kind are based on correlative studies that are as distant from the phenomena as are the manifold theories of time proposed by philosophers since the time of St. Augustine. As we see later, most psychologists propose explanations based on the impact of other cognitive functions such as short-term memory or attention.

The reason for this lack of coherent theory is that philosophers and psychologists share the same problem in dealing with phenomena of temporal cognition; we are still not sure what psychological time is, what its properties are, or even how it can be objectively measured. Even more conspicuous by their absence are any robust theories of the cognitive mechanisms of time perception. All we really have are general descriptions of its phenomenology (e.g., "time goes faster when one is busy"). Where theories and explanations are offered, they are of a dubious nature and fluctuate from theorist to theorist without adding deep understanding, much less of ultimate resolution.

How bizarre, strange, and counter to the laws and rules of physical reality time perception is, is clearly illustrated by a recently observed and relatively macroscopic phenomenon: *change blindness*.[16] Change blindness occurs when environmental discontinuities are completely ignored by the perceptual system. Such a proclivity on the part of the human observer tends to smooth over substantial changes that may have occurred in the stimulus scene. As a result, astonishingly drastic changes in the environment can be completely ignored by the observer. This can be as dramatic as an observer's inability to see that one person completely replaced another if the scene is momentarily interrupted (Simons and Levin, 1998), the sudden disappearance or appearance of relatively large objects in a visual scene, or even heads being exchanged in photographs. All that is required is that there is a momentary interruption of the scene for the observer to be blind to the physical fact that a major discontinuity in the scene had occurred.

Initially, change blindness was attributed to the interruption of the scene by a blink or eye movements (Grimes, 1996; McConkie and Currie, 1996). Subsequent research, however, has shown that virtually any kind of interruption of the scene can result in an observer being oblivious to fact that what was a major portion of the scene had changed. This interruption can be in the form a scene switch in a motion picture (Levin and Simons, 1997) or the passage of an opaque object in front of a background—a background that changed during the obstructed vision interval (Simons and Levin, 1998).[17] Indeed, even a brief flicker interruption of about 80 msec can inhibit the detection of a difference between two scenes on either side of the discontinuity (Resink, O'Regan, and Clark, 1997).

The facts of change blindness have been succinctly summarized by Simons and Ambinder (2005) as follows:

1. Change blindness occurs whenever attention is diverted from the change signal.
2. Changes to objects that are central to the meaning of the scene or changes in visually distinctive objects are detected more readily than other changes.
3. Attention may be necessary for change detection, with changes to unattended objects going unnoticed.
4. Attention to a changing object may not be sufficient for change detection (p. 45)

However, even these generalities seem to underplay the compelling and robust nature of the change blindness phenomenon. Nowhere is this better illustrated than the famous gorilla experiment reported by Simons and Chabris (1999). Here, in a manner analogous to the change blindness experiments just discussed, observers were oblivious to dynamic and unexpected additions to their visual environment when specifically attending to other aspects of the scene. This phenomenon, therefore, has also been designated "inattention blindness."

In Simons and Chabris's experiment, observers viewed a video of a scene of two teams of players, one in white shirts and the other in black shirts, passing a basketball back and forth. The observers were instructed that they were to count the number of passes between players on one or the other of the two teams. However, in the middle of the video, an unexpected event occurred; a person dressed in a full-body gorilla suit walked across the scene. The results of this experiment were astonishing; depending on the conditions of the experiment (the color of the shirts, the difficulty of the attention-demanding task, etc.), approximately half of the observers in the experiment never reported the presence of the totally unexpected gorilla!

This work on change and inattention blindness has attracted a good bit of attention and publicity since it was rejuvenated in the 1990s. As fascinating as this phenomenon is to the lay public and as challenging as it is to psychological theorists, no solid explanatory theory has been forthcoming. Instead, the general "explanation" offered for this cognitive process is in terms of the interference from other cognitive processes, some defined and some only vaguely alluded to. Pearson and Schaefer (2005), for example, carried out experiments that compared the meaningfulness and the centrality of interest as well as the degree of cognitive engagement. Their rather obvious conclusion that "change blindness is not solely a data-driven perceptual phenomenon" (p.1451) illustrates both the disconnect between the actual physical stimulus and the perceptual phenomenon and the difficulty that all psychologists have when asked to explain how perceptual phenomena arise. Indeed, none of these verbal models actually provide any insights into the mechanisms of this phenomenon.

Other possible and seductively simple explanations such as the limited capacity of short-term memory, raise more questions than they answer. For example, if we are able to process (store and access) only a small portion of incoming information, how do we reconstruct a usable "model" of the external world? I doubt that such questions will ever be answered. The best we can do is to behave like behaviorists and describe and determine the necessary and sufficient conditions to elicit these fascinating illusions.

3.5 INTERIM SUMMARY

This chapter has reviewed what is, at best, an incomplete sample of perceptual phenomena in which the properties of space and time in the cognitive world appear to be different from those in the physical world. All of these phenomena are characterized by discrepancies, nonveridicalities, or distortions between the information provided by the stimulus and the way in which that information is interpreted or reconstructed in our experience.

We have seen how paradoxes of time and space seem to operate by laws that are quite different from those of physics. Thus, even the most fundamental requirements of scientific deduction, such as the requirement that a cause must precede an effect do not always appear to hold in the mental domain. Other basic properties necessary for quantification, namely ordinality and equal intervals, are regularly violated in mental events. Psychological time is shown to be elastic, stretching in accord with other cognitive activities, and sensitive to its environment in a way that defies the concept of strict or even orderly causal relationships between stimuli and responses. If this argument is correct, it would have profound implications. Even a limited behaviorist goal of defining stable transformations between stimuli and responses would become problematic.

In other more recently examined instances, there is an extraordinary insensitivity to discontinuities in the smooth flow of time. Catastrophic discontinuity, unbecoming to an orderly world, seems to characterize our ability to switch between alternative interpretations of ambiguous stimuli. All in all, the world of the mind is irregular and disorderly, missing some of the order and regularities that made it possible for physics to prosper.

Although there is little question that these discrepancies between physical properties and psychological properties exist, because of the disorder and irregularity, it is difficult to explain them using the standard methods of axiomatic-deductive science. Indeed, it may be that because of the inaccessibility of mental processes and the differences between the basic properties of physics and psychology, we may never be able to do more than describe these phenomena. Part of the fragmentary nature of psychophysical research may, therefore, be explained in terms of the lack of generalizable relations from one phenomenon to another.

These discrepancies and nonveridicalities are a major part of the argument that mental processes are virtually inexplicable. If the laws of science of the outside world are not followed in the mental world, and if we do not have adequate measurement and access to the inside world, no conceivable means of inferring internal processes and mechanisms from behavior is possible.

Certainly, as I have noted throughout this chapter, there is no current convergence on any explanation of any of these phenomena. Neuroreductive explanations vie with cognitive explanations; yet, both must contend with the weakest possible explanation of all: merely linking one phenomenon with another in an unpromising effort to bring some order and understanding to the major scientific and philosophical problems considered here.

A most important idea, however, is that, whatever explanations that may be provided in the future, valid or not, it really does not matter. The essential fact is that the laws, processes, and rules that seem to govern cognitive processes are functionally different from those regulating physical events. Smythies (2003) put it succinctly when he alluded to one of the properties of consciousness:

> the demonstration by recent experiments in neuroscience and psychophysics that we do not perceive the world as it actually is but as the brain computes it most probably to be. (p. 48)

Although Smythies goes on to suggest that there are three kinds of "ontologically independent" realities (space-time, matter, and consciousness), I don't think that such an extension of dualistic thinking is necessary. Modern physics and brain-mind, matter-function hypothesis conceptually link the three "realities" into a single kind of conceptual schema. Unfortunately, no one has yet convinced anyone of the "true" ontology, and there is a high probability that such questions (among many other intractable natural phenomena) are beyond the scope of reductive science. I leave that knotty problem to others.

What is undeniable is that this chapter repeatedly illustrates that the rules and properties of the world of physical time-space are different from those of conscious experience, and there is no way to bridge the chasm that separates them. This means that there is nothing currently (and perhaps never can be) comparable to the Cosmological Principle of physics, which provides stability and uniformity to the cosmological and quantum sciences. Thus, that which is inaccessible in the mind cannot be inferred in the same manner as are the nature and properties of physically inaccessible objects and events.

One result of the absence of a psychological equivalent to the Cosmological Principle, as well as the indeterminacy of behavioral observations, is that the mathematical techniques appropriate for physics are not likely to work for psychological phenomena. It is to this topic I turn in the next chapter.

NOTES

[1] Relativistic physics also leads to many so-called paradoxes. However, all such paradoxes eventually turn out to be mistaken intuitive interpretations of events that are difficult to make conform to the world of the human scale. Furthermore, those paradoxes only begin to emerge at relativistic speeds near the velocity of light. In this chapter, all of the mental, cognitive, and psychological "paradoxes" occur at our scales of time and space.

[2] The light cone is the ensemble of events that have in the past influenced a current event and can be influenced in the future by a current event. Since a cause or influence must be within a distance determined by the speed of light at any time, the cone converges on the instantaneous present from the past and diverges from it in the future. This restriction determines the biconical shape of the universe of events that can interact with each other.

[3] It is important to appreciate that the paradoxical reversal of cause and effect of the metacontrast phenomenon is quite different from the so-called "paradox" inherent in asking a person "not to think" about something (e.g., Wegner, Schneider, Carter, and White, 1987). In that case, the paradox is that an observer's conscious efforts to "not think" about something are typically followed by an increase, rather than a decrease, in thinking about that very thing. There is no reversal of temporal causality as there is in the examples of apparent motion and metacontrast. There are a host of other such "paradoxes" in psychology that do not represent violations of time or space in the sense I use here, only the persistent uncontrollability of human thought.

[4] Even physical theory is not final. Recent developments have challenged this concept by proposing that gravity, like the photons conveying electromagnetic forces, is conveyed by particles called gravitons. Unfortunately, not even the footprints of gravitons have yet been detected. We must be careful to distinguish between descriptive mathematical theories and physical entities. Many of the former have not yet been associated with the latter. One of the most notable lacunas in linking formal

models to physical structure is the success of string theory in describing the behavior of physical processes and events and the lack of any physical evidence for the existence of strings.

[5]Many previous "neural" theories of this simultaneous contrast illusion have been based on presumed lateral inhibitory interactions among neurons in the nervous system. Although such theories are not implausible for edge effects such as the Mach Band, they seem inapplicable here because of the global (overall) change in lightness of the gray areas.

[6]Indeed, one interesting aspect of illusions is the source of their continuing popularity in introductory textbooks and in salons of all kinds, despite this theoretical impenetrability. I suppose it is because illusions startle and amuse us by their unpredictable distortions that few of us really care about explaining or applying them. It is only in the rarest incidences that they are of any practical importance. One important exception is in the dangerous illusions that occur in flying, especially in landing an airplane. Gibb (2007) describes how distortions of space can lead pilots to misperceive the correct glide path in such a task and, all too often, fly into the ground rather than gracefully landing.

[7]It might be useful at this point to interject a brief reminder. Our inability to go from the neural substrate to the phenomenological responses (as indicated by our verbal behavior) is primarily a practical problem introduced by complexity and the sheer size of the numbers involved. However unlikely such a neural synthesis might be because of the great complexity of the neural networks and the resulting intractability of the mathematical problems involved, it is not in principle impossible. On the other hand, our inability to go from the behavioral utterance to the underlying cognitive processes that account for the discrepancies between the stimuli and the response is an in principle barrier that can never be overcome, no matter how profound the logic and how powerful the instruments. The reasons for this are manifold, but not the least of them is that reports of cognitive phenomena do not convey sufficient information for reductive explanations; that is, they are underdetermined. This means that there are many equally plausible explanations for every problem that cannot be empirically distinguished: the "one-to-many" problem.

[8]I am indebted to a comprehensive review on the McCollough effect by Humphrey and Goodale (1998) for calling this quotation to my attention.

[9]The full story of the evolution of color theory, especially of the eventual compromise between the Young- Helmholtz trichromatic and the Hering opponent color theories, can be found in my earlier books (Uttal, 1973; 1981). For the moment, the important message is that the psychophysical data from color mixing and neutral points supporting opponency, although suggestive, is not definitive. It conflicts with equally compelling trichromatic theories of sensory transmission. The answers emerging when the salient neurophysiology was carried out was that both encoding schemes were used at different levels of the ascending pathways. This is one of the cases in which the conflict between indeterminate color phenomena could be and was eventually resolved by neurophysiological research. A similar direct means of transforming the indeterminate into the determinate is not available for higher cognitive processes such as the illusions of time and space that we are dealing with here.

[10]Recent efforts (Huk, Ress, and Heeger, 2001) to use fMRI images to demonstrate aftereffects in the medial temporal cortex (visual area 5) suggest that there may be such a response at the cumulative population level.

[11]It is interesting to note that although the actual complexities of these aftereffects have been well known for over a century, the myth of fatigued neurons persists. Furthermore, recent discussions of the very important McCollough contingent aftereffect, so contrary to the persistent, but apparently erroneous zeitgeist, argue against the fatigued neuron hypothesis. Nevertheless, this evidence to the contrary has almost disappeared from introductory textbooks and the fatigued neuron myth persists in our classrooms.

[12]An excellent history of the historical efforts to define and to explain time can be found in what may have been the most insightful book on the topic ever written: Fraisse (1963). In his introduction, he spells out the complexity and uncertainty that has beset studies of psychological time for millennia. Unfortunately, most of the difficult questions about time that he poses remain unanswered. Excellent historical reviews of the debate over discrete and continuous psychological moments in time perception can be found in VanRullen and Koch (2003) and in Sacks (2004).

[13]I should also note at this point that speculation begins to take over from empirical findings. Neumann and Nieple (2004) conclude their interesting review by noting that although, in general, we do have a fairly veridical "correspondence between perceived temporal order and the order of events in the physical world" (p.263), this may be due to some compensatory factor of which we know very little. Indeed, they offer the equally mysterious process of "time constancy" as a possible explanation of the difference between the orderliness of reaction-time measurements and the apparent disorder of temporal-order judgments. This common, but deeply flawed, strategy of explaining one phenomenon by another speculative one does not satisfy the need for serious scientific explanation.

[14]The three categories of time to which Fraisse (1963) alluded in this context were intervals less than .5 second; intervals from .5 sec to 1.0 sec; and those more than approximately 1 sec. They do not exactly overlap with Woodrow's, but the point being made is the same; subjective time seems to obey different rules for different interval sizes.

[15]The shape of this sliding window is not constant but is something like a Gaussian distribution. It efficacy is low when it first approaches a point in time; it increases to a maximum as it passes by, and decreases as it flows on always in what appears to be a continuous flow.

[16]Although change blindness in real world scenes became of theoretical interest only in recent years, some earlier studies are, in retrospect, now appreciated to be simple exemplars of it. DiLollo (1980), for example, observed a primitive form of the phenomenon in which observers could not tell that two dot arrays separated by a brief interval were different over wide ranges of dot density.

[17]Resink (2002) comprehensively reviews the many different kinds of interruptions that can lead to the change blindness phenomena.

4

Statistics and Mathematics in Psychology and Physics[1]

4.1 INTRODUCTION

Psychology can hardly be considered to be an atheoretical science. Psychologists constantly "explain" the observed behavioral phenomena by neural, cognitive, and mathematical models and theories of a wide variety. For many reasons, however, most of these theoretical approaches have failed to provide robust and compelling explanations of the observations. Psychological theories typically remain constricted to narrowly defined fields, in many cases dealing only with the results of single experiments or circumscribed sets of phenomena. Certainly, no one claims that a comprehensive theory covering many psychological phenomena comparable to quantum mechanics or relativity has yet emerged from this plethora of microtheories.[2]

Some of the most widely advertised psychological "theories" simply don't achieve their goals. There is wide agreement that we have no explanation of how sentient mind emerges from neural activity. Nor is there much hope on the horizon, as I have previously argued (Uttal, 2005), that neural theories are likely to provide a bridge between the mind and the brain, although this argument is often disputed, especially in popular publications. Similarly, many "cognitive" models are useless hand waving involving verbal restatements or redescriptions of the phenomena rather than reductive explanatory explanations.

There is, on the other hand, a highly regarded field of "mathematical psychology" that has had considerable success in representing some aspects of behavior. However, as I discussed in Chapter 1, even the most distinguished contributors to this field (e.g., Luce, 1995) have raised questions about the validity of "mathematical psychology" and the justification for its existence; no argument, however, from me. It remains a highly respected approach to theory-building in psychology.

There is, however, a major distinction between the mathematical formulation of most psychological theories, on the one hand, and most physical ones, on the other. In the main, physical phenomena are described by the deductive mathematics of number theory, algebra, calculus, differential and integral equations, and a number of other specialized fields such as topology. These powerful axiomatic-deductive tools have been collectively referred to as mathematical analysis; they are referred to hereafter in this book as conventional mathematics, or sometimes simply as analysis. Occasionally, some physical problems involve so many interacting objects that statistical concepts and methods are invoked, the most familiar example being the statistical mechanics of gases. This approach can be used in situations in which the individual particles are essentially identical and the forces operating between them relatively simple and homogeneous. In the main, however, most physical theories are framed in the methods and terminology of conventional mathematical analysis.

Mathematical psychology, on the other hand, is most often characterized by statistical methods, both of a descriptive and inferential nature. In a few and relatively rare instances, usually involving sensory or motor responses, conventional analytical mathematics may be used. An increasingly familiar example is the application of dynamical systems theory to motor skills. In the main, however, most theories of learning, choice, personality, intelligence, and other cognitive processes are framed in statistical terms.

The distinctly different choices of the most appropriate mathematical tool for physics and psychology, respectively, illustrate a major difference between the two fields and one that speaks directly to the main thesis of this book: that psychological inaccessibility is different from physical inaccessibility in a way that precludes valid inferences from behavior to the underlying mental and/or neural mechanisms and processes. The root of this problem, as I have argued so far in this book, is that there is nothing comparable to the Cosmological Principle (which permits us to assume that the laws of physics are the same everywhere) in the psychological domain. As a result, we have no basis on which to make the parallel assumption that the laws of our physical world and the laws of our mental world are the same.

This chapter explores the reasons for this difference and examines the forces that have driven psychologists toward statistical methods and approaches and physicists towards analytic ones, respectively. The ultimate outcome of this exploration is an appreciation that the methodological differences result from the fact that many of the propositions and properties of the physical world do not conform to those of the mental world. Analysis and physics fit together, and statistics and psychology fit together, respectively, both because of the nature of the mathematical tools and the properties of the respective subject matters.

I am convinced that the differences between the two approaches are not trivial or simple, but are understandable in terms of quantitative differences arising from differences in their respective phenomena of interest. I argue here that their theories and applicable mathematical methods differ because the subject matters differ in the extent to which they are complex, not in any qualitative differences in their basic nature. In other words, the argument made here is that physics and psychology are ontologically identical (i.e., both are manifestations of an underlying materialism), but require different epistemological strategies to study. I suggest that the choices of their respective formal approaches are, therefore, reasonable and natural responses to those differences.

Let's begin by examining some of the properties of analytical mathematics and statistics that make for good conceptual fits to the properties of physics and psychology, respectively. Physical theory is inherently deductive. This fits well with the tendency of physical laws to emerge from derivations from simpler axioms, Newton's and Einstein's models both being excellent examples of this process. Simple axiomatic statements lead inexorably to certain predictions by a process of deduction. The forces are identifiable and, even more important, are relatively few in number. In ideal physical situations (e.g., mechanics) the nature of the forces can be inferred from the course of the derivation, at least in the form of a description of their effects.

On the other hand, statistics is inherently inductive simply because it is designed to deal with processes that are intrinsically complex and situations in which the effective forces are not always identifiable. In most cognitive studies, for example, there are many unidentified forces simultaneously at work that require careful control of the experimental design. Thus, whereas deductive mathematical analysis is driven by axioms, derivations, and a few initial conditions, inductive statistics is mainly driven by multifactorial data. Where analysis seeks to (and can often) trace the course of a process, statistics seeks to consolidate a mass of observations without peering deeply into the processes and mechanisms that account for those data.

Physics fits well with deductive analysis (and vice versa) because physical systems are relatively simple and the forces involved are almost always specifically identified or at least potentially identifiable. A major result of this simplicity and identification of forces and variables is that theories typically pyramid in physics. We can start with something like Newton's three laws as axiomatic postulates and derive the detailed and precise behavior of entities such as the solar system. The deductions (i.e., theorems) then suggest more general laws and principles that can be consolidated with others that may initially seem to be unrelated. The major physical forces are identified, and others can be considered to be minor perturbations. In physics, the expectation is gradually being fulfilled that a universal theory uniting all physical forces is not only possible but imminent.

Statistics, on the other hand, deals with the world as it comes—a complex and inseparable aggregate of many forces. In an analogous fashion, psychologists typically fail to unravel the many forces and processes that produce a cluster of observations. It is rare in cognitive psychology, furthermore, for all of the effective forces to be controlled, much less identified, and, unlike physics, many of them may not even be potentially identifiable. Psychology uses statistics to infer and describe properties of the mental world for which no hope of axiomatic reduction exists. Indeed, the very idea of a unified theory of psychological behavior that transcends a limited range of phenomena seems outlandish from a current point of view.

Analysis and statistics also differ in their respective goals. In the main, physics attempts to account for and predict the specific behavior of a specific object or the collective behavior of a uniform set of objects interacting under the influence of a few simple laws. For example, a physicist might be interested in the trajectory of a projectile, the pressure of a gas, or the forces holding the atomic nucleus together. For the mathematical physicist, every case in a repeated sample of observations must follow the same basic laws previously enunciated. A single violation would be the undoing of an entire theoretical approach; a single erroneous step in a derivation could destroy a complex logical construction and predict nonsense. Although the mathematical physicist must be able to deal with variability, in some cases (e.g., gas dynamics) proof of a theorem usually depends on a much higher level of confidence (i.e., many more standard deviations) than that usually asked of behavioral scientists. Whereas psychologists are willing to accept .05 (approximately two standard deviations) as a criterion for a difference between two conditions, physicists demand at least three standard deviations, and five and six standard deviations are considered desirable levels of "proof" in basic particle physics.

Conventional mathematical analysis, furthermore, does not deal well with uncertainty in its outcomes even though there may be uncertainties in its measurements.[3] If one applies the rules properly, then after agreement has been reached on the initial axioms and states, anyone who deductively searches for an answer should come to the same conclusion. Indeed, this is the nature of a proof in mathematics: an irrevocable, convincing derivation from an axiomatic starting point to an inevitable final state.

However, statistics is very tolerant of variability. If fact, it dotes on it. Thus, there are major differences between statistics and its ancestral source—analytic mathematics—in goals and strategies as well as in methodology. A deductive analysis may be valid regardless of the context. In fact, it can be effectively argued that it is its very generality that makes it so useful. In some pure cases, we may go so far as to say that mathematical theorems may be totally independent of any observations or measurements. Only after many years may a pure mathematical derivation turn out to have some entirely unexpected application.

Statistical manipulations, on the contrary, are meaningful only to the degree in which they are embedded in some context. It is meaningless to say that the average value is "7" in the absence of a context; it is perfectly meaningful to discuss the continuity of a mathematical function which may not currently represent some physical process.

Hall (2006), in a Power Point introduction to her statistics course, summarizes the different ways that mathematicians and statisticians deal with context as well as anyone I have encountered:

Mathematical Thinking

- Mathematicians rely on context for motivation and for sources of problems for research.
- The ultimate focus of mathematical thinking is on abstract patterns.
- Context is part of the detail that must be "boiled off" to reveal the true structure.

Statistical Thinking

- Statisticians look for patterns, but whether the patterns have meaning and value depends on how these patterns interweave with the story line.
- In data analysis, context provides meaning.

The inductive foundation of statistics means that its major goal, almost by definition, is the description and evaluation of an aggregated set of data defined in a particular context. Statistics, thus, has two roles to play. The first is that it provides an efficient method for the accumulation and description of what may be an extensive body of data from a limited universe of study. The implicit goal is to make patterns visible that may be invisible in the uncondensed mass of observations. This is the role of descriptive statistics. Accumulated data are condensed to a small set of descriptive values, such as the mean, standard deviation, and range, in order to summarize the trends in the uncondensed data and, thus, to describe the behavior of a complex system. This is a role of convenience; an obscured pattern may be made apparent, a central tendency of a variable pattern of responses determined. However, these numbers, as useful as they may be, are meaningful only within the context under study. The value of the variance in one context means nothing in another context. Numbers, expressions, rules, and theorems in conventional mathematics, on the other hand, must be universally meaningful and independent of the context.

The second role of statistics is interpretive. This is the role of inferential statistics. The task in this case is to estimate and describe the properties of a universe that extends beyond the sample of data that had been accumulated. Inferential statistics depends on some very important assumptions, not all of which are understood by users of this kind of mathematics. The most general one is that the sample of data that has been measured is assumed to be characteristic of all of the measurements that might have been made.

Furthermore, hidden away in the mechanics of statistical thinking is the idea that the various observations are independent of each other. That is, there must be no influence of one observation on others. For example, there should be no sequential dependency; each measure should be independent of those that preceded it. If there is, the data are assumed to exhibit a kind of bias, and the observations may not represent the original universe from which the sample was drawn. In some instances, there may be a priori assumptions about the expected nature of the data that may influence the way they are processed. There are a number of other more specific assumptions (e.g., normality of the distributions) that are embedded in many statistical operations that are not required of analysis.

The result of this foundation of explicit and implicit assumptions is that the meaning of a statistical analysis depends on the task or context at hand, whereas mathematical analysis may be totally independent of it. There is, therefore, a vast difference between what is an acceptable outcome in analysis and statistics. We summarize some of these properties in the following list:

- Mathematical analyses require a level of certainty, specificity, and constancy that statistics does not. Statistics suggests general relationships; conventional mathematics demands specific ones.
- Statistics accepts and describes variability; mathematics does not suffer variability gladly. Statistics always deals with response probabilities, therefore, uncertainty. The deductive mechanisms of analysis demand certainty, although many applications and measurements may involve, or at least tolerate, some variation when one compares predictions and measurements in the real world.
- The meaning of a statistic depends entirely on the application; analysis can thrive without an immediate application.[4]
- Not only is statistics inductive, but it also depends on certain assumptions that are not necessary for analytic mathematics. For example, statistics assumes that a sample is an accurate representative of the original population. Furthermore, it assumes that the sample is taken from a population that is reasonably stable and does not change during the course of an experiment. Considering the adaptability of human behavior, it is possible that this is one of the great misunderstandings of psychological research. Physics, to a much greater extent, can deal with isolated and functionally stable events, such as the trajectory of an individual particle or meteorite.
- Statistics is susceptible to a variety of different types of error or biases. The sample size may be too small to permit valid inferences; the population may be impermanent or changing; the inferences drawn may be distorted or blatantly wrong, among many others. From an epistemological point of view, the very meaning of knowledge in conventional mathematics and physics differs.

Notwithstanding the many differences between analytic mathematics and statistics, I am not suggesting that statistics is not a part of mathematics, or that either one is exclusive of the other. The rules of arithmetic must apply to statistics in the same way they apply to conventional mathematics. Statistics is, without doubt, a branch of mathematics, and its methods, symbols, processes, and procedures cannot violate mathematical rules. Furthermore, nothing is absolute. There is no sharp line between the two domains; there is a continuum from subject matters that are appropriate for analysis and those that are more suitable for statistics.

Therefore, we have to be careful not to overstate the limits of either statistics or analytic mathematics or the differences between them. There are many common ways in which statistical and analytic mathematicians operate. Statisticians are very concerned with the application and the processing of real world numerical data. On the other hand, although pure mathematicians do not always concern themselves with real life situations, applied mathematicians, not to mention physical scientists of all kinds, are often concerned not only with the logical elegances but also how well their mathematics represents the universe under study. Whatever the proclivities, mathematical rigor is demanded at all levels of their activities. This rigor demands that the rules and procedures of mathematics be inviolate from one context to another, regardless of the presence or absence of an application.

Rather, the argument made here is that there are pressures from the subject matters themselves that drive one kind of science to analysis and another to statistics. Hand (2004) summarized the differences in subject matter between physics and psychology in a particularly succinct and insightful manner, after citing Wigner (1960), who said:

> if there were no phenomena which are independent of all but a manageably small set of conditions, physics would be impossible. (p.4)

Hand, apropos of this insight, went on to say:

> In psychology, quite the opposite is true: essentially all variable[s] are related and the trick is to tease apart this complex tangle. Whereas in physics *all* electrons are *identical*, in psychology *no* two people are identical. (p. 152)[5]

Stigler (1999) had a similar take on the situation. Physicists, he noted, differ from psychologists in some profound ways. For physicists, any variability in their data is assumed to be a perturbation, a minor fluctuation that is to be superimposed on the deterministic main relationships. Psychology and other sciences with similar challenges, on the other hand, concentrate on that variability and only rarely assume that there is a simple, unique cause explaining the data. I enjoyed Stigler's (1999) clear statement concerning the physical perspective:

> When an astronomer resorted to statistics in the 1820's, and the tool he usually reached for was the method of least squares, there was no doubt in his mind that he was after something real, definite, objective, something with an independent reality outside of his observations, a genuine Platonic reality inherited from the then unshakable evidence of Newtonian theory. (p. 190)

Psychology, on the other hand, has no foundation on which to build such a "Platonic reality." It deals with intangible thoughts that have no mass, velocity, or any of the other properties of a real or "independent reality." Not only are these physical properties missing, but time itself is not constant or irreversible (i.e., it is not isotropic, homogenous, continuous, or monotonic, as discussed in Chapter 3). Furthermore, the observations collected in a typical psychological experiment are terribly variable. Thus, response variability becomes the key issue in place of objective reality. Stigler (1999) concluded by also pointing out:

> Astronomy could exploit a theory exterior to their observations, a theory that defined an object for their inference.... Experimental psychologists could, through experimental design, create a baseline for measurement, and control the factors important for their investigation. (p. 198)

The differences between psychological and physical strategies, therefore, grow out of the basic fact that physics deals with at most a few interacting forces that can be associated with an objective reality. There are few psychological phenomena in which this kind of simplicity is to be found. The much more usual case is that a collection of several or many poorly identified forces are driving the behavioral outcome of what are, at best, only hypothetical mental processes. The physicist's approach collapses when more than a "few" interacting forces are at work. Then the problems become intractable (e.g., consider the persistent failure to solve the general three-body problem of classical physics). Psychology and the other social sciences face this problem all of the time and finesse it by measuring the probabilities of occurrence of the various possible outcomes.

Thus, there are differing goals, methods, and properties of mathematical physics and statistical psychology. Physics seeks and, to a degree, has succeeded in determining the underlying nature of even those parts of the universe that are not directly accessible. This was accomplished because of the constraint provided by the Cosmological Principle. Although some psychologists believe that their quest is of the same kind, and that they are searching for the basic laws and properties of an equally inaccessible human mind, some more thoughtful scholars have pointed out that psychology should not and cannot (if the behavioral view is correct) hope for the same level of accomplishment. For example, Hand (2004) reminds us of the wise counsel of Anne Anastasi, who pointed out for a specific case—factor analysis—what should be a watchword of psychological thinking in general:

> Factor analysis is no longer regarded as a means of searching for *the* basic, fixed, universal units of behavior. Rather, it is recognized as a method for organizing empirical data into useful categories through an analysis of behavioral uniformities. Like the test scores and other observational data from which they are derived, factors are descriptive, not explanatory. They do not represent underlying causal entities. (p. 134)

This might well be said of any statistical model of human mentation.

It is for these reasons—the differences in the nature of psychological and physical phenomena—that the mathematical tools appropriate for each field of science have diverged. To understand the impact of the subject matter on the choice of method, it is useful to provide a brief set of definitions of some basic concepts. The next section serves this purpose.

4.2 A MINILEXICON OF RELEVANT TERMINOLOGY

Some of the general statements made in the introduction to this chapter can be clarified by more precise and specific definitions of some of the critical terms. However, it should be reemphasized that the two sets of definitions I present here—those pertaining to conventional mathematics and those pertaining to statistics—are not totally exclusive.

The differences between the two fields, therefore, are matters of approach and emphasis. Nevertheless, even these differences help to explain both the different properties of psychological and physical phenomena and the particular advantages or disadvantages of using one kind of mathematics or the other to measure those properties. The purpose of this section is to clarify the kind of thinking in a way that may lead one field of inquiry to choose conventional mathematics and another to use statistics.

4.2.1 Mathematics in General

The history of mathematics virtually overlaps the history of civilization. It emerged from an early appreciation of the concept of counting, that things had a value in terms of their numerousness as well in terms of their physical properties. We don't really know when mathematics started, but there is reliable paleoanthropological evidence that people started counting tens of thousands of years ago. Inscribed stones and bones from the upper Paleolithic period indicate that things were being counted long before they were being described in writing. Much later, in the Bronze Age, tokens indicating the number of sheep or containers of grain came into widespread use in Sumeria and Egypt. In fact, it is thought by many archeologists (for example, Dreyer, 1998; 1999) that these simple tokens may have been the precursor of written languages in these two early civilizations.

There have been many conceptual steps, some complex and some simple, from these early enumerations to modern mathematics that were critical to its current highly developed state. The emerging awareness of quantity separate from the objects themselves, the invention of number systems, carrying, and the discovery of zero are among the many possible critical concepts on which this powerful tool was founded.

Among a few of the other most notable contributors to this long history are:

- Anonymous Babylonians who seemed to have solved right triangles (the so-called "Pythagorean" problem) and even anticipated the solution of quadratic equations.
- Thales, perhaps the first "scientist" in the modern sense of the word, who introduced some of the Egyptian geometrical ideas into Greek thinking.
- Pythagoras, the grandfather of modern mathematics, famous for his application of mathematics to all facets of existence.
- Zeno and his paradoxes in which one never reaches goals.
- Archimedes, a major contributor to early ideas of geometry.
- Euclid's geometry, as published in his great book "Elements."
- Anonymous Chinese who invented the decimal system and negative numbers.
- Anonymous Arabs and Indians who preserved mathematical thinking for over a century through medieval times following the decline of Greek and Roman civilizations.

From the Renaissance on, the list of contributors to mathematics is too great to fully detail here. A few of the most notable great mathematicians of the past are Fibonacci, Copernicus, Galileo, Napier, Fermat, Kepler, Pascal, Newton, Bernoulli, Leibniz, Euler, Fourier, Galois, Cauchy, Maxwell, Hilbert, among many others.[6]

Despite this long history, mathematics remains almost as difficult to define as mind and has many different meanings for many different people. A good place to start is with a typical dictionary definition. One of the most succinct is:

> The study of the measurement, properties, and relationships of quantities and sets, using numbers and symbols. (*The American Heritage® Dictionary of the English Language*: Fourth Edition. 2000)

Another dictionary-type definition is:

> Mathematics is the body of knowledge centered on such concepts as quantity, structure, space and change. ...Mathematics explores such concepts, aiming to formulate new conjectures and establish their truth by rigorous deduction from appropriately chosen axions and definitions.[7] (Wikepedia: The Free Encyclopedia)

Other efforts to define mathematics have attempted to do so by exhaustively enumerating the many different kinds of mathematical methods, tools, and subfields. However, this kind of exhaustive definition finesses one previously mentioned aspect that would otherwise help us enormously to focus on the essential meaning of the term *mathematics*; that is, analytical mathematics can survive without any context or data or application. Mathematics, in its purest sense, requires only internal consistency; pure mathematics can eschew any relevance to worldly matters and simply deal with the logic of symbol manipulation. All that is required is that rules be established for the manipulation of abstract symbols. Mathematics is also the mother science of measurement and quantification from which we derive the ability to manipulate symbols that do represent numerical values.

At its most basic level, however, mathematics is number free and can function perfectly well with only a few basic a priori logical assumptions and relationships. This was the contribution of Whitehead and Russell's (1910–1913) great work: *Principia Mathematica*. They were able to provide a firm foundation for mathematics based on a small number of logical assumptions and the acceptance of a few rules. None of their definitions required that the symbols had any attachment to physical events and objects or even to numerousness. Implication, rather than addition, characterized their heroic effort to seek the logical roots of mathematics.

Although the purest forms of mathematics do not require a link to physical reality, this does not mean that they may not have unexpected physical applications. Some of mathematics' most important contributions have arisen from its eventual application to worldly matters. A curious fact is that over the course of the history of applied mathematics, there has been a convergence of its purely logical origins and the properties of the physical world. Mathematics often develops abstract methods that only later turn out to have relevance or are useful in a particular application. Of special interest to this book is the anticipatory development of non-Euclidean mathematics in advance of its application to relativity theory (see pages 8–9).

Despite what are often purely logical origins, one of the most extraordinary attributes of mathematics is how well it describes the physical world. This is the domain it currently serves so well. Why this fit between mathematics and the physical world exists is not immediately obvious. Mathematics' millennial-old roots are founded in simple counting, and numerousness is a physical attribute, so it must have been influenced by the nature of the world in which it originated and then evolved. However, there are other possibilities. Whitehead and Russell's major contribution was to show that mathematics of many different kinds could emerge from purely logical foundations without any connection to physical concepts.

The mysterious origins of mathematics raise several possible theories of why the material world and mathematics fit together so well. One possibility is that the physical world is also orderly, logical, and regular, and it was just an accident that it happened to fit equivalent properties of mathematics. If this is the case, then the compatibility of physics and pure mathematics represents a coincidence of cosmic proportions. Wigner (1960), a thoughtful commentator, alluded to this point when he said:

> the enormous usefulness of mathematics in the natural sciences is something bordering on the mysterious and there is no rational explanation for it.... it is just this uncanny usefulness of mathematical concepts that raises the question of the uniqueness of our physical theories. (pp. 1–2)

The other major possibility, however, is that mathematics evolved in the natural world as we know it to fit its properties to those of that world. In a different world, other kinds of mathematics might have evolved. The one we have, however, is the result of the natural forces extant in our world, directing us toward a relevant form even though these forces may have been cryptic or unappreciated by the founding fathers of a particular mathematics. If this explanation is correct, it becomes understandable why conventional mathematics should not fit other domains with differing properties—such as psychology.

Another important and general attribute of analytic mathematics, in particular, is its pyramidal structure. Classical Egyptian, Babylonian, and especially Greek mathematics are as useful today as they were when first formulated. Indeed, the fundamentals of arithmetic and geometry known to Pythagoras and Euclid are still taught in our schools as prerequisites to the equally persistent seventeenth century developments such as the calculus, all of which provide the foundation for even more modern mathematical tools. A wonderful property of mathematics, therefore, is that the newest and most complex ideas are built on the foundation of the past. This facilitates modern science, as a new theoretical foundation does not have to be constructed for every new observation.

As I have already noted, this kind of pyramiding is not something that is characteristic of modern psychological science. In psychology, theoretical pyramiding is notably rare or absent, and new theories emerge for almost every new observation. Psychology, therefore, requires something a little different, a little more appropriate for handling a higher degree of variability, and a little more tailored to its needs. Something that is more inductive than deductive. Something called statistics.

4.2.3 Statistics

As ancient as mathematics is, one of its daughter sciences—statistics—is very modern. Therefore, its specific history is very well known, and many of the major contributors have been recognized since their seminal works.

The history of statistical thinking goes back only to the seventeenth century, when the first concepts of probability began to emerge. Statistics was forced into existence by the need to solve problems that involved large amounts of variability, multiple measurements, and, at best, incomplete or indeterminate data. No longer could one count on the dependability of a single object, event, observation or measure to establish knowledge; instead, it was necessary to accumulate and estimate the nature of a universe of discourse that was only partially known from a subset of all of the possible observations. This was quite a different approach from that required for the much simpler behaviors of typical physical systems.

Some of the most important personages and conceptual steps in the history of statistics were:

- Christiaan Huygens (1629–1695), who, influenced by Newton, published the first work on mathematical probability.
- Archibald Pitcairne (1652–1713), a physician who may well have been the first applied statistician as he applied mathematical ideas to medical problems.

- John Craig (1667–1731), the first "statistical" social scientist who studied the probability of historical events.
- Abraham de Moivre (?–1754), the actual inventor of the misattributed normal or "Gaussian" distribution.
- Thomas Bayes (1702–1761) has usually been credited with being the first to use the idea of probability as a tool in inductive thinking.
- Pierre-Simon Laplace (1749–1827), to whom is usually attributed the first use of such properties of a distribution as the mean and standard deviation.
- Friedrich Wilhelm Bessel (1784–1846), an otherwise already well-known mathematician, who developed a statistical procedure for combining the differing observations of astronomers in measuring stellar transits—the so-called "personal equation."[8] (An important historical fact of considerable interest is that the application of statistics to the social sciences, although sporadic and infrequent prior to the early nineteenth century, was actually antedated by this application in astronomy.)
- Johann Carl Friedrich Gauss (1777–1855), whose work resulted in his famous least squares method for predicting the most probable outcome of a multitude of observations.
- Francis Galton (1822–1911), who inaugurated the concept of the "regression to the mean," among other statistical ideas.
- Ludwig Boltzmann (1844–1906), the mathematical physicist primarily responsible for the creation of statistical mechanics, a methodology useful in the study of heat and gaseous behavior.
- Karl Pearson (1857–1936), who contributed the idea of tests of significance, most notably the chi-square test.
- George Udny Yule (1871–1951), who was the most influential statistician in reminding us that correlation is not causation.
- Ronald A. Fisher (1890–1962), who introduced the idea of random design in agricultural statistics and ultimately in psychological experimental design.

Again, I must remind my readers that this list is a very small sample of the many important contributors to statistical and pre-statistical thinking. For those who would like to delve deeper, there is nothing better than the wonderful and insightful discussions of statistical history presented in Stigler (1999).

What, then, is this relatively young subfield of mathematics all about? Dictionary definitions of statistics comparable to those offered for mathematics include:

> The mathematics of the collection, organization, and interpretation of numerical data, especially the analysis of population characteristics by inference from sampling. (*The American Heritage Dictionary of the English Language*: Fourth Edition, 2000)

and

> Statistics is the science and practice of developing knowledge through the use of empirical data expressed in quantitative form. It is based on statistical theory which is a branch of applied mathematics. Within statistical theory, randomness and uncertainty are modeled by probability theory. Because one aim of statistics is to produce the "best" information from available data, some authors consider statistics a branch of decision theory. (Wikipedia, the Free Encyclopedia)

Other less comprehensive definitions just refer to statistics as the "branch of mathematics that collects, analyzes, and interprets data." I believe this simplistic approach to a definition to be incomplete, in that it does not distinguish the role of statistics from that of many other kinds of mathematics.

Perhaps statistics can be simply described as an applied branch of mathematics that is especially concerned with probability or uncertainty produced by incomplete data sets.[9] In general, as mentioned previously, statistics is a field of mathematics that depends heavily on what always turns out to be variable data. Its main goals are to summarize variable and sampled data and then to attempt to infer from those summaries, something about the universe from which the sample was drawn. It is a means of processing and drawing conclusions about an incomplete data set when the outcomes occur only with certain probabilities, probabilities that cannot be estimated until data are accumulated.[10]

Probability. Probability is defined as a measure of the likelihood of an event. By likelihood I mean the relative proportion of times that the specific event occurs from among all possible events that could occur. Probability is usually measured as a fraction of 1. If a particular event always occurs, it is assigned a probability of 1.0. If a random or uncertain event occurs unpredictably often but, say, one out of ten times, then it would have a probability of 0.1, and so on. An event that could never occur would have a probability of 0.0.

If a particular event occurs with only a particular probability, this implies that the specific outcome is uncertain in advance of its occurrence and cannot be predicted with certainty. This uncertainty may be introduced into a situation for one of two reasons. There may be unknown factors that "randomly" influence the outcome of a number of possible events that cannot be known, no matter how often we repeat an observation or how precisely we make our measurements. An example of such a situation is dice casting. Atmospheric conditions, slight imperfections in the dice themselves, the original forces involved in casting the dice, and the surface on which they are thrown make it, for all practical purposes, impossible to predict how they land. This explanation of uncertainty asserts that as a result of immeasurable forces, it is fundamental and unknowable.

Uncertainty may alternatively arise, not because of any fundamentally unknowable factors, but simply because we do not yet know enough about the situation. This may be rectified, to an unknown extent, by the collection of an additional amount of data. However, as with any inductive process, we can never exclude the possibility that an improbable event will occur next that will shatter our expectations, our axioms, and our theories. Even then, although we may have much to say about the distribution of probabilities of an event, we may not be able to predict the behavior of a specific individual in the sample or the possibility of a specific event. Probability thus implies a certain degree of uncertainty and unpredictability of individual cases or events. Statistics is the mathematics of choice to handle this kind of environment.

Randomness. The subfield of mathematics known as statistics has evolved to handle both the actually immeasurable and the currently unknown sources of randomness. Randomness is a key concept in statistics but extremely difficult to define. Randomness is sometimes equated with unpredictability or the absence of a pattern, but this conflates the cause with the effect. Nonrandom patterns may be subtle and undetectable by the means available to us. In psychological experiments, it usually is desired to have some source of random numbers in order to "counterbalance the experimental trials" to assure that some undetectable bias is not introduced into the study by sequence effects.[11]

Testing for randomness is a serious matter in many fields of science because a conclusion may vary depending on whether or not a sequence was truly random. Despite this importance, it is not always easy to determine whether a number is truly random. The classic Chaitin-Kolmogorov incompleteness algorithm, for example, states that a string (usually of bits) is random if and only if it is shorter than any computer program that can produce that string. Since one can never completely test a sequence to see if it meets this criterion, it is now generally agreed that no sequence can be rigorously proven to be random. Rigorous Chaitin-Kolmogorov randomness is sometimes distinguished from a less severe form of "statistical randomness," which simply requires that a sequence is presumed to be random if no pattern can be discerned by any "reasonable method."

Although the issue of randomness is rarely explicitly considered in psychological experiments, the possibility of bias, to the point of invalidating the experiment's conclusion, is ever present, even when a pattern cannot be "discerned." That a pattern is not recognized does not mean that a subtle pattern is not present. In point of fact, we just do not know what the impact of randomness is in many of our experiments, in spite of a wide variety of statistical tests for randomness of short sequences (see the original work of Kendall and Babington Smith, 1938).

4.3 WHY IS PSYCHOLOGY STATISTICAL?

Now that we have spelled out some of the properties of mathematics and statistics, we can turn to the major question confronted in this chapter: Why is quantitative psychology predominantly statistical in its theories and models? There are several answers to this question that should already have begun to become evident. In this section, I hope to clarify these reasons in a way that makes it clear that statistical methods and thinking are appropriate responses for the science of psychology, whereas something different is usually required in the physical sciences.

The simplest and most direct answer to this question is that the properties of psychological responses map onto the properties of statistics better than onto conventional mathematical analysis. Statistics is characterized by its attention to and its ability to process such properties as variability, incompleteness, indeterminativeness, probability, uncertainty, and randomness. Beyond these formal properties, statistical analysis also represents an approach to quantified measurement and thinking that seeks to accumulate observations, to summarize them, and to infer the true nature of a complex mass of findings, which cannot always be studied in their entirety and which are individually meaningless. It is, from this point of view, an inductive strategy. It is quite different from the deterministic goal of fitting functions and graphs of relatively invariable and complete observations that is characteristic of analytic mathematics, the latter being a deductive, or more completely, an axiomatic-deductive strategy.

Thus, when we step back a bit from the frenetic pace of our daily lives in the laboratory, it becomes obvious that psychological responses exhibit many of the same kind of properties that characterize statistics. From the outset of any deep consideration of their nature, it is immediately obvious that psychological observations are, at best, extremely variable. Anyone who has spent any time in a perceptual or cognitive laboratory knows that an individual response tells us virtually nothing about the process under study. Furthermore, every "trial" is totally indeterminate and virtually incapable of providing even a starting point for the drawing of conclusions. The simplest questions (for example: Did you see that?) may be responded to in ways that are totally meaningless in the context of the single response. Untrained observers may not have understood the question, confused about what "that" was, their attention may have been distracted, or some other cognitive process (e.g., shyness) may have interfered, penetrated, or misdirected the decision processes underlying the desired response. A serious additional problem is that observers in experiments may not be playing the "game" intended by the experimenter because of poor training, inadequate instructions, ignorance, stupidity, or even downright meanness. All of these influences lead to a high level of variability in psychological research.

Unlike its counterparts in physics,[12] any single behavioral response is incomplete and uninformative to the extreme. It becomes useful only as a "sample" of a much larger population of responses. The entire mass of psychophysical and cognitive research methodology is designed to control some of the irrelevant factors just mentioned. Furthermore, its goal is to overcome the indeterminativeness of the individual response by collecting multiple responses and statistically accumulating them in order to estimate a more realistic (i.e., more probable) value of some observed behavior. From this point of view, psychology and statistics mate well because one is designed to handle variability and incompleteness, and the other is variable and incomplete. Physics and analysis mate well because both share the common property of a high level of completeness and relatively low levels of variability.

The ultimate source of the indeterminativeness and incompleteness in a psychological experiment is that the responses of humans are variable as well as indeterminate. Whatever the source of response variability—fundamental uncertainty or ignorance—it is clear that there is little hope in the foreseeable future that psychology ever will graduate from statistical to more conventional forms of deductive mathematical analysis.[13] Therefore, the theories emerging from our research are likely to be phrased in terms of probability for as far into the future as we can predict.

This conclusion, if correct, has some important consequences. One of the most important is that it is never (and I use the word "never" intentionally) going to be possible to predict the results of individual experimental trials, much less individual human behavior! The best we can hope for is to sharpen our estimated probabilities of a particular behavioral outcome by greatly increasing the number of behavioral responses collected. This raises the problem of the economical allocation of resources and the practical limits of data collection resources. Even if we had had unlimited resources, not all analyses scale up to increased precision with increased sample size; some saturate or break down completely as more samples or components are added. Thus, what may be significant results with small samples may lose their significance when larger samples are used. "Proofs" of the existence of parapsychological phenomena are particularly susceptible to rejection when sample sizes are increased.

The implication of this persistent uncertainty for the present discussion is that the fit between statistics and human psychology (and other sciences that deal with uncertain subject matters and variable or partial data) is not temporary. We are going to have to live with it and its advantages and disadvantages permanently.

In summary, for a variety or reasons, there is now a better fit between the properties of psychology and statistics than between those of psychology and conventional mathematical analysis. It is likely that this state of affairs is going to continue since it reflects fundamental properties of both psychology and statistics. The answer to our rhetorical question at the outset of this section, therefore, is that psychology and statistics are drawn together because they are so much alike.

4.4 INEVITABLE STATISTICAL OUTCOMES AND THE OBFUSCATION OF EMPIRICAL RESULTS.

Deductive mathematical reasoning represents to many the epitome of logical analysis and power. It is hard to imagine how many of the most important scientific and technological developments of the past could have occurred without the development and application of such tools. However, mathematics, like any other tool, is not perfect. It has its limits. For example, it is not always appreciated during the development of a reductive theory that its mathematical formalisms are not themselves intrinsically reductive or explanatory. Mathematical analysis makes obvious what was only implicit in the data and thus may direct us to plausible new experiments. However, discovery (or, at least, confirmation of predictions), more often than not, results from the observation of natural events. Occasionally, a new phenomenon is pointed to by the deductive process, but in the main, it is the empirical study of nature that leads to novel and unexpected discoveries. Mathematics is, therefore, neutral with regard to internal processes and mechanisms; it describes process but cannot in principle uniquely identify the underlying causes accounting for the behavior of mind or matter.

In place of reductive explanations, mathematics provides us with a powerful means of describing data and even predicting future events. Even the best mathematics is not able to discriminate between analogous mechanisms and processes or to distinguish between a multitude of different, though plausible, ways that a particular behavioral outcome might be instantiated in neural or cognitive processes or mechanisms. There are a number of different arguments for these constraints on the reductive powers of mathematics. The next sections describe a few of them.

4.4.1 Mathematical Fictions

One of the most overlooked limits of mathematics as an explanatory or reductive tool, despite its practical utility, is that sometimes mathematical procedures inject their own properties into the understanding of a problem. That is, they can often create fictional properties and attach them to models and theories. These fictional properties have a seductive tendency to be reified as real rather than metaphorical entities. This is especially true for psychology in which the objects of study are typically abstractions or hypothetical constructs (MacCorquodale and Meehl, 1948), created to instantiate what on closer examination is nothing more or less than a description of the results of experiments. Thus, an observed process—for example, a change in behavior—becomes an entity designated as "short-term memory," or the results of a mathematical analysis become "frequency channels."

In another earlier work (Uttal, 2003), I extensively discussed a number of these intrusive effects in which physical fictions were created by the application of a mathematical procedure. I now list a few, along with an example or two to illustrate this point.

Fourier Analysis: The mathematical reduction of a function to a set of sinusoids can lead to the fiction of spatial frequency channels in the visual system.

Polynomial Curve Fitting: Equations such as

$$R = a_0 + a_1\phi + a_2\phi^2 + a_3\phi^3 + \ldots + a_k\phi_k \qquad \text{Equation 4.1}$$

where ϕ is almost any kind of function are so general that they can always fit a function and thus provide a perfect theory (i.e., perfectly fit some data). The problem in this case is that there are an unlimited number of degrees of freedom (Nimh, 1976).

Pseudomathematics of Intractable Formulae. Equations such as

$$I_0 = \Sigma_k^n \cdots\cdots \Sigma_l^n N_{K\ldots} \cdots\cdots N_{l\ldots} \qquad \text{Equation 4.2}$$

proposed by Rashevsky (1948) to represent the "entire organic world" cannot be solved because they are too complex to be evaluated. Rashevsky himself stated that it was insolvable because it "involves lots of organisms, lots of parameters and lots of values for these parameters." The numerical value of "lots of" can only be guessed at.

Zipf's Law. Zipf (1935) observed that the rank order (r) f words (defined by how often they were used) and the actual frequency (F) with which they were used was well fit by the power law expression:

$$F \infty \frac{1}{r^k} \qquad \text{Equation 4.3}$$

where k is an exponent that is very close to 1 and ∞ means "is closely approximated by." He extrapolated from this incontestably valid empirical expression to develop a theory of the underlying "Psycho-Biology" of language production based on what he referred to as the "Principle of Least Effort." Unfortunately, this law is now known to describe many natural phenomena and has nothing special to do with "Psycho- Biology."

1/f Noise. A particular kind of noise distribution was originally thought to reflect a variety of psychological properties. 1/f processes were observed in timing operations by Gilden and his colleagues (Gilden, Thornton, and Mallon, 1995; Gilden 1997; 2001); in choice experiments by Clayton and Frey (1995); in an apparent motion experiment by Brady, Bex, and Frederickson (1997); in serial interactions in pronunciation reaction times by Van Orden, Moreno, and Holden (2003); and in motor reaction times by Wagenmakers, Farrell, and Ratcliff (2004). It is now known to be a general property of systems in which stimulus forces are accumulated or pooled to produce responses.

A number of mathematicians have proposed possible explanations for the ubiquitous 1/f function. For example, Bak, Tang, and Wiesenfeld (1987) proposed that it emerged whenever dynamical systems with sufficient numbers of degrees of spatial freedom self-organize toward critical states. These critical stable points or states were, according to them, typically characterized by 1/f functions. These stable points are further characterized by scale invariance (i.e., it works at all sizes) and emerge when minimally stable entities interact. The implication of their analysis is that it is the process of self-organization rather than the properties of the individual systems that leads to the 1/f distribution.[14] Mandelbrot (1999) also linked the ubiquitous 1/f phenomenon to the property known as "self-affinity," which he believed was analogous to the scale invariance of fractal dimensions. Some plausible, although unproven, suggestions about its origins in psychological studies include:

- Hausdorff and Peng (1996), Pressing (1999), and Gilden (1997) all suggested that 1/f function appeared as a result of the summation of cognitive functions that have different time scales.
- Pressing and Jolley-Rogers (1997) argued that it was the result of a combination of cognitive and motor components and their interaction with memory.
- Busby and Townsend (2001) suggest that the 1/f effect is caused by variations in the extent to which the observer is attending to the stimuli.

Criticism that the 1/f function was psychologically significant was summed up by Hausdorff and Peng (1996):

> This suggests the possibility that complex fluctuations and 1/f scaling observed in many biological systems does not reflect anything "special" about the mechanisms generating these dynamics. (p. 1)

and by Wagenmakers, Farrell, and Ratcliff (2004) who stated that:

> It appears that the presence of LRD [long-range dependence or 1/f functions] per se cannot be used to uniquely identify a specific underlying psychological process or structure. (p. 608)

4.4.2 Information Pooled Is Information Lost

In the previous section, I dealt with mathematical or statistical outcomes that can add spurious meaning to data beyond that which is determined by the nature of the data themselves. There is another source of distortion when one applies statistical procedures, even when they are intended to be nothing more than descriptive. That source is summed up in the expression:

INFORMATION POOLED IS INFORMATION LOST!

These five words encapsulate one of the most prevalent misunderstandings in psychological research. To begin to understand why this is so, let us review a bit of the logic behind the application of statistics to psychological phenomena.

Statistics is a powerful tool for describing and summarizing highly variable phenomena exemplified by those associated with human behavior. Because no single observation can definitively establish the expected nature of a behavior, we find it useful to collect a large number of observations in order to suggest average, typical, or group values. From these measures of central tendency, it is possible to predict the probability, but not the certainty, of a particular response. To preserve all of the details of all of the observed responses, however, would require that we tabulate each of them individually. This complete tabulation would not only be tedious but the volume of the resulting data set itself is likely to obscure the very pattern of probabilities being sought.

To overcome these practical difficulties, psychology and many related sciences have developed and regularly use means of compressing or consolidating the observations by pooling or accumulating them in one manner or another. This process of accumulation is the essential practical heart of statistical analysis. It is a powerful tool, it is a convenient tool, and it allows us to make judgments about the average nature of the phenomena in which we are interested. It does not, however, permit us to determine underlying causal relations or the neural or cognitive processes and mechanisms that instantiate the observed behavior.

To pool data in the way usually done is to lose information. It does not matter how cumbersome it would be to handle the full bulk of original observations; whenever we combine things there is a price to be paid. Just as we cannot unscramble an egg, there is no way to reconstruct the original data set from estimates, for example, of their median or mean value. In fact, once the pooling has been done, there is literally an infinite number of combinations of data that can produce that same average value.[15]

This loss of information has profound implications for a broad range of psychological theories. Most important, it means essentially that predictions of individual responses from statistical summaries are not possible. Although this may seem an arcane point, the inability to interpret past behavior or predict future behavior is a major problem for many fields of psychology that is not generally appreciated, particularly in contexts such as the courtroom or in the burgeoning field of cognitive neuroscience.

Thus, the price paid for statistical summation is a profound loss of the detailed information contained in the original data set. No matter how necessary or convenient or helpful such pooling may be, no matter how well it consolidates the observations, it leaves us in the dark about the individual datum and precludes specific predictions or estimates of individual behavior in the past or in the future. All that we have left is some probabilities.

Furthermore, statistical measures not only lose information about the individual responses, but also obscure such basic information as how many responses had been made, unless additional efforts are made to maintain counts. A measure of central tendency such as the mean or median, by itself, is an expression that is independent of sample size. Since sample size is critical in defining the utility of a statistical analysis (e.g., in determining its level of significance), here, too, is a price paid for the conveniences provided by statistical summaries.

This loss of information is not limited only to the kind of behavioral data sets that the psychologists deal with, but also pertains to much of the work in current neurophysiology. Researchers who use field neuroelectric phenomena (such as

EEGs or evoked brain potentials) to study the relationship between brain activity and cognitive processes often assert that these signals contain "all" of the information that is present in the individual neuronal responses that contribute to them. However, this is incorrect. The pooling of individual neuronal responses to produce a field of electrical activity, exemplified by the EEG, also throws away virtually all of the information inherent in the individual cellular responses. To be sure, the reduction of the data may be useful, but it is incorrect to assume that the molar EEG contains "all of the information of the neural network from which it emerged." Instead, in order to provide a measure of brain activity that fits an available technology, the electroencephalographer surrenders most of the information concerning the interactive activity of the individual neurons. Unfortunately, it is at this level that mental processes are most likely to be encoded.

4.4.3 Emergent Psychological "Laws"

Another way in which data analysis can distort the realities of behavioral variability occurs when inevitable mathematical processes and data pooling combine to produce a fallacious law supposedly describing the nature of a behavioral process. This is especially prevalent when mathematical expressions or descriptive laws are used to summarize the outcome of psychophysical experiments. Such summary expressions are extremely seductive; however, there is increasing evidence that they may represent distortions of human behavior by implying a kind of uniformity that may not exist in any psychological or biological sense. Thus, for example, when the Hick-Hyman (Hick, 1952; Hyman, 1953) law relating the reaction time RT and information content of an array of stimuli (in terms of the number N of alternative responses) is expressed as:

$$RT = K \log_2 N \qquad \text{Equation 4.4}$$

a misleading impression is created of greater precision than actually exists.

The reason for this conceptual error is based on the fact that the RT is not an absolute function of N in the usual mathematical sense; in other words, the "=" sign is a fiction. Instead, in the psychological context, "=" should be interpreted not as a determinist and quantitative equality, but rather as a symbol for a much softer kind of relationship, something more akin to "is generally related to." The reaction time for a single event is not strictly predicted by this equation. All we can really assert is that only that, on a statistical basis, the reaction time tends to be prolonged by many alternative stimuli (N) and reduced when there are only a few. The duration of the individual reaction time is unpredictable. Furthermore, the sign "=" suggests to the uninitiated that "$K \log_2 N$" is the sole influence or cause in determining the RT. This is obviously incorrect; the implication is based on a false conceptual confusion of the kind of laws that exist in psychology with those of physics.

Other putative "laws" of psychological function are often expressed as standard functions, implying the relation of the full range of responses (R) to the full range of the stimuli (S), with equations of the form R = f(S). In addition to the standard problem of an imprecise meaning of the relationship "=", there are other difficulties with such simplistic relationships, relationships that can hardly be called "laws" in the same breath as those of physics. One of these additional problems is that extended data sets may be equally well expressed by exponential or power functions, hyperbolic functions, or logarithmic functions, none of which are unique, and all of which may be closely related to each other in either shape or mathematical roots.

There are, therefore, three main sources of difficulty in the establishment of a mathematical law describing, for example, the relationship between stimulus intensities and subjective magnitudes:

- The various laws are often so much alike that it becomes very difficult to distinguish whether one or another best fits the averaged data. Very precise criteria and assumptions must be fulfilled to warrant the choice of one "law" over another, and the best-fitting function is not always the most useful theoretically. Simplistic criteria such as how well a theoretical expression "fits" the data may be meaningless in a context of a glut of free variables. Roberts and Pashler (2000) have pointed out that a good theory not only fits the data, but also identifies those that are implausible or unlikely. The excellent fit obtained with such expressions as Zipf's Law conceals the fact that any cognitive or neural hypotheses associated with it may be wildly irrelevant. Since mathematical expressions are neutral with regard to underlying mechanisms and processes, any physiological or structural assumptions are also independent of the mathematical ones. Many different neural or cognitive processes may account for the same behavior.
- In some cases, different laws are duals of each other in the sense that one derives naturally from the other. Thus, the averaging of a collection of one kind of function may produce a summary function of another type. For example, integrating hyperbolic functions can lead to the production of logarithmic functions (Killeen, 1994), and power laws can emerge from averaging of exponential, range-limited logarithmic, or even range-limited linear curves (Anderson, 2001; Anderson and Tweney, 1997). Thus, pooling data may suggest properties for the averaged responses that are not present in the individual responses. This distortion has been well known for many years, as evidenced in articles by Sidman (1952), Bakan (1954), and Estes (1956). Estes (2002), in particular, identifies some empirical tests that demonstrate the distorting effect that were carried out by Anderson and Tweney (1997) and by Heathcote, Brown, and Mewhort (2000). Thus, the process of averaging can mistakenly produce a particular type of function, and even worse, a particular theoretical orientation that may have no relevance to the matter under study.

The point is that the properties of a putative law may be an outcome of the accumulation process, rather than of the cognitive phenomenon under study. This does not negate the fact that the nature of the data may also influence the outcome. It just says that there are a number of factors, some substantive and some procedural, that may affect the choice of a summary law when data are pooled other than the data itself.

- Some "laws" fit a pooled average of data in parts of their range better than in others. For example, logarithmic functions typically do not do as well with low values of the parameters than with high values where the plot is compressed. The generality of the "law" to represent the data over its full range is thus open to question.

Furthermore, we must not overlook the ever-present fact that even the best fitting laws simply may not be adequate approximations to the data. It is no exaggeration to assert that no "law" of psychology has ever been found to exhibit the precision characteristic of Newton's three laws.

Examples of how misleading a psychological law may be can be found throughout classic psychophysics. Perhaps the most criticized of all was the first of all, Fechner's (1860/1966) logarithmic law:

$$S = k \log(I - I_0) \qquad \text{Equation 4.5}$$

in which S is the subjective magnitude of an experience, k is a constant, I is the physical intensity of the germane stimulus, and I_0 is the threshold stimulus intensity. For many years it has been appreciated that this is, at best, a crude and imprecise approximation. However, it was not until Stevens (1971a) proposed a power law alternative to the logarithmic law in the form:

$$s = k(I - I_0)^n \qquad \text{Equation 4.6}$$

In this expression, n can take on a wide range of different values, maximizing as large as 3 for the effects of direct electrical stimulation to the teeth, and as small as .03 for the brightness of a point source of light.

Stevens' formulation had some significant advantages over Fechner's Law in that it is much more general; it could represent functions that grew "exponentially" when n was greater than 1.0, approximated linear functions when n was equal to 1.0, as well fit those that approached asymptotes when n was less than 1.0. Nevertheless, there were only slight differences between the power function and the logarithmic function over wide ranges of $n < 1$.

However much of an improvement it may have been, a number of criticisms have been raised about the power law as a general theory of human sensory performance. Many of these criticisms have been based on the now familiar idea that power laws may emerge when data is averaged as a natural result of the pooling process, rather than from any property of the stimulus or the response. A power law, although widely accepted throughout psychology, may be another of those functions that inevitably emerge whenever data representing the interactions within complex nonlinear systems are combined.

An additional suggestion that the power law is not a valid indication of cognitive reality is implicit in its broad applicability outside of the domain of psychology. Power laws similar to Stevens' formulation appear regularly throughout many different scientific enterprises, as reviewed by Schroeder (1991).[16] A specific recent example of a power law can be found in the work of Vandermeer and Perfecto (2006) concerning the population size of agricultural pests. They indicate that the power law is the result of pooling "independent populations growing exponentially" (p. 1001). The implication of their theoretical insight, once again, is that it is the cumulative process at work, rather than a direct property of the insect's biology that produces the power function.

Empirical studies in psychology have been concerned with this problem since the early work of Jones and Marcus (1961) and Green and Luce (1974). These researchers were among the first to appreciate that individual psychophysical observations were described by power laws only when pooled and not in their raw form. Indeed, although Stevens (1971b) himself was concerned with the problem, he apparently did not realize the ubiquitous nature of the power function throughout so many other areas of science—a fact that should have suggested to him that his formula might have also been an artifact of the pooling process. Stevens did, however, consider the problems that might emerge with different kinds of averaging. Interestingly, he suggested that:

> When error grows in proportion to magnitude, so that the relative error stays constant, a logarithmic transformation tends to undo the skewness. (p. 441)

This may be interpreted as implicitly reflecting his appreciation of the possible distortion introduced by various kinds of transformations. Whether he extrapolated this difficulty to the power function he so influentially introduced in psychological thinking is not known.

Empirical studies continue to challenge the validity of the power law as a descriptor of this psychobiological reality. Recently, Heathcote, Brown, and Mewhort (2000), in a heroic review that looked at a wide range of data from 24 published studies, discovered that power functions did not do as well as exponential functions in representing the unpooled results of learning experiments. Furthermore, they were able to develop a combined law based on exponential functions that was more accurate in representing data than was a power function for pooled data.

In a more formal and less empirical manner, the power law has also been criticized because of what were thought to be flaws in its basic assumptions. Narens (1996) approached the problem of the validity of Stevens' formulation by identifying the two key assumptions he believed were "inherent in his [Stevens'] ideas about ratio magnitude estimation" (p.109). I have paraphrased these two implicit assumptions as:

- There is a ratio scale type function that an observer in an experiment uses to measure the subjective intensity produced by a stimulus.
- Each stimulus can be used as a modulus [reference], so that a functional relationship can be developed between the range of stimuli and the range of subjective intensities. In other words, the numbers assigned by the observers are used to produce the function describing the relationship between the stimulus values and a mathematical function.

Narens went on to observe that these two assumptions imply that the responses by the subject should be multiplicatively related to each other. Indeed, he points out that this second property "imposes powerful constraints on the subject's magnitude estimation behavior" (p. 110). If the multiplicative property is not adhered to by the data collected with the magnitude estimation method, Narens argued, then the entire power function approach may be flawed. Testing for multiplicativity thus becomes a critical element in validating the magnitude estimation procedure and eventually the power law formulation.

Narens asserted that the key experiments to justify this kingpin of psychological theory have never been carried out. He also pointed out that one of the biggest flaws in the entire field of psychology, as well as in the power law approach specifically, is that there is no a priori reason to assume that what an observer says is a true reflection of the intensity of the experience.[17] He concluded:

> Stevens and other magnitude theorists do not provide any theoretical or even intuitive rationale for this [assumption]; at most they only note that the method of magnitude estimation produces representations (that they presume to be part of ratio scales) that interrelate in consistent and theoretically interesting ways with other phenomena. It is my conjecture that the consistency results not by reflecting some underlying reality but from a lack of enough relevant data that might reveal structural inconsistencies. (p. 110)[18]

If we combine this failure of the magnitude estimation theory to match the properties of the data with the idea that power functions can emerge from a variety of other underlying functions when data are pooled, the entire approach of asking subjects to act as introspective measuring instruments in this manner becomes questionable. Given that a power law may describe a data set for reasons other than it reflects the data (i.e., because of the generality of the expression and because of its emergence from the summation of non-power law components), it is clear that its universality is based on some highly tenuous mathematics, assumptions, and leaps of scientific faith.

The problem is further exacerbated by certain logical problems. Both the logarithmic law and the power law, on reflection, are much too simple. They imply a simple one-dimensional relationship between the stimulus intensity and the perceptual response. Given the complexity of human cognitive processes and well-known influence of other factors (for example, sequential interactions and even something as fundamental as dark adaptation), it seems simple to the point of naiveté to assume that such a direct causal relation exists between our percepts and the numbers we use to express them. The unspoken but implicit admonition that such a law holds only if "all other factors are held constant" does not mitigate this problem. Clearly, all other effects of the stimulus and the cognitive state of the observer are not held constant in any conceivable psychological experiment. Any subjective magnitude estimate is the outcome of many other factors than simply the physics of the stimulus. Memorial factors, sequence effects, contrast effects, and other properties of the stimulus (such as its onset time and time course) may have far more profound effects than is usually appreciated when a power law is suggested as a summary of a set of data.

4.4.4 How Emergent Laws Hide Variability and Other Properties

Above and beyond the extent to which the so-called "laws" of cognitive processing validly represent phenomena, there is a further problem in their application. All of these "laws" from Fechner on are mainly expressions used in two ways. The first is that they are often used as a basis for suggesting underlying mechanisms; that is, as clues to what processes might be explaining their particular shape. The second way they can be used is to approximately represent general tendencies and as a convenient way of organizing data. From one point of view, this distinction is equivalent to the stress between cognitivism and behaviorism, respectively. Let's consider each approach separately.

The most familiar example of the first—approximate use of a so-called "lawful relation" to develop explanatory theory—was Weber's original law for the increasing size of the just noticeable difference of stimulus intensity (ΔI) with stimulus intensity I.

$$\Delta I = kI \qquad \text{Equation 4.7}$$

This early law of psychophysics asserted that there was a specific relationship between ΔI and I; the larger the difference between two stimuli had to be before they were perceived as being unequal. This simple fact led Fechner to propose the logarithmic law of the growth of stimulus intensity and Stevens to propose, as an alternative, the power law. Thus, both of these two nonlinear relationships do roughly reflect some of the intensive aspects of human perception and perhaps some other processes, such as the effect of practice. "How rough?" is an empirical as well as a mathematical problem; however, there is no question that a basic nonlinearity is a property of psychophysical judgments, and that Weber's Law captures this relationship generally if not specifically.

An important theoretical question concerns the locus of this nonlinearity. In an earlier work (Uttal, 1973), I reviewed a large body of evidence that contributed to our understanding of its source. Clearly, as already noted, there was a nonlinear and generally compressed relationship between the stimulus and the observer's response. Neurophysiological data showed that this nonlinearity also existed between the stimulus and the response of the receptor cells of the various modalities.

However, there appeared to be an approximately linear relationship between the output of the receptor cells and the parameters of the afferent neuron's responses. Although the data were fairly rough, it also appeared, therefore, that there was a more or less linear relationship between the higher order afferent neurons and the psychophysical responses.

The logarithmic or power law nonlinear relationship between the stimulus and the perceptual response, therefore, was accounted for, to a very large extent, by the transductive properties of the receptors. Indeed, if you bypassed the receptors and stimulated the afferent neurons directly with electrical currents, the sensory relationship to the stimulus was best described by a power law exponent greater than 1.0![19]

Data such as these made it clear that our entire sensory systems are not linear in any psychobiological sense and gave some credibility to the general concept of nonlinear laws as valid reflections of the real biology of our cognitive systems. This is not being denied. What is being suggested is that although these laws are generally true, prior to direct neurophysiological experimentation, they were inadequate clues to a satisfactory theoretical explanation. These descriptive laws, therefore, were and remain neutral to reductive explanation.

The second way that these expressions are used is to conveniently display data and suggest approximate relationships between stimuli and perception. This nonreductive, atheoretical, descriptive approach attributes nothing to the neurophysiology of the systems under study. Instead, it is nothing more than a graphic expedient. Logarithmic curves, in particular, have the highly useful property of compressing data, so that the full dynamic range of some variable can be shown on a convenient-size graph. Higher values of a variable are demagnified, so that they take up less room on a graph than lower values and vice versa. This graphical tool for data presentation sometimes mirrors some of the properties of the psychological response. However, this graphical convenience is not identical in meaning or concept to the compression of the dynamic range expressed in Weber's Law; one is only a data analysis tool and the other is a behavioral reflection of a complex of neurobiological responses. It is almost a pun or coincidence that the graphical tool should have a similar form as the response.[20]

There is, however, a massive price to be paid for this convenience. All of these compressed and expanded laws act to obscure the data and to suggest simplicity and order where they may not be present. One example is the use of log-log coordinates in which both axes are compressed in this manner and can be used to present wide ranges of data without requiring huge plotting papers. However, a substantial amount of information is lost when this convenience is exercised. The list of lost information includes the following (Tufte, 1983, was our guide in developing this list):

- Variability of the data in highly compressed regions may be hidden. Thus, unequal variability over the entire range of values may be obscured.

- The size of the intervals between equal numerical steps changes at different logarithmic values. High values are crowded; low are more widely separated than they should be.
- The true skewness of a relationship is obscured because of the compression of higher values or the expansion of lower values.
- Essential nonlinearity may be hidden by the fact that many log-log graphs appear to be linear.
- The intercepts on a log-log graph may be distorted since the logarithm of 1.0 is 0.0.

Of course, any transformation has the same problems. Power laws compress and expand depending on the exponent. Coordinate systems can be transformed from Cartesian to Polar forms or to phase spaces. Each has some useful attributes distorting or obscuring some trends and enhancing others.

In general, however useful these transformations may be in emphasizing one feature of a function or another, they can also obscure or mislead us about some other aspect of the relationship between a stimulus and a response. The general point to keep in mind is that because of variability, randomness of responses, the resulting need to pool psychological data, and the inevitable properties that some mathematical functions may impose, precision in psychophysical laws is elusive. If these difficulties are confounded with the even more fundamental difficulties of measurement in psychology introduced by the scarcity of ratio scales, and thus the absence of equal intervals and non-arbitrary zeros, it is obvious that there are many, many problems that occur when mathematical (including statistical) methods are applied to the study of psychological phenomena.

Many of these problems remain beyond the ken of most current psychologists. The elite group of mathematical psychologists that appreciates and understands the role of mathematical thinking in psychology is relatively small. One can only hope that a deeper awareness of what we actually are doing when we use models and formulae to represent cognitive processes eventually percolates down into the general research community.

4.4.5 On Correlation and Causality

Statistics at its best can show that there is a tendency for functions or processes to be related to each other. However, as Yule (1926), among many others, has pointed out:

CORRELATION DOES NOT IMPLY CAUSALITY!

Like mathematics in general, statistical correlations are neutral with regard to the causal forces or factors at work not only in complex systems, but even in simple systems.

There is considerable misunderstanding about the meaning of a correlation and the extent that we can draw conclusions about underlying causes from it. Famously, a recent poll showed that 64 percent of the general population believes that correlation *does* imply causation. Considering the subtle meaning of these two words, this outcome may not be too surprising—for the general population. However, it is shocking to observe the same fallacy being regularly expressed in cognitive neuroscience circles. It is a mainstay of arguments in that field that *correlated* neural signals are the ipso facto *equivalents* of related psychophysical responses. There is perhaps no other area of science in which mere concomitancy or covariation is so wildly misinterpreted as a causal relationship.

There are many famous examples of how correlations may be confused with causal relationships. Some of them are obviously ridiculous (e.g., "shark attacks and ice cream consumption are correlated") and lead to the bad reputation that statistics has among the lay public (e.g., "statistics can be used to prove anything"). However, many subtle correlations can be equally misleading to members of the research or education communities (e.g., "experience studying Latin and taking IQ tests is correlated").

Not all psychologists commit the logical error of imputing causation to correlation in such a simplistic fashion. Some, most notably Cheng (1997), appreciated the limits of correlation as a source of causation, but sought to finesse the problem for relatively simple situations. Cheng clearly stated the problem as follows:

> Because causal relations are neither observable nor deducible, they must be induced from observable events. The two dominant approaches to the psychology of causal induction—the covariation approach and the causal power approach—are each crippled by fundamental problems. (p. 367)

In her extensive review of the problem, therefore, both of these methods are deemed to be flawed; the covariational approach because:

> all covariation models of causality face a major common hurdle: As many have noted, covariation does not always imply causation. ΔP [the contingency between candidate cause and effect] is clearly insufficient as a criterion for causal induction, because not all covariational relations are perceived as causal. Many things follow one another regularly, yet one does not infer a causal relation between them. (p. 367)

and the causal power approach, which depends on a priori knowledge or intuitions, because:

> First, it suffers from the weakness of not being computational: It does not explicitly define a mapping between the ultimate input to the causal induction process and its output. Proponents of this view have not explained how the domain-independent knowledge that all events are caused can constrain inference from covariation in a specific domain. Second, with respect to the problem of causal induction, the interpretation of the Kantian view in terms of specific causal powers is crippled by its circularity. Specific causal powers are, by definition, causal. (p.368)

Although Cheng went on to propose a solution to overcome these problems (based on a combination of the covariation and causal power approaches, her solution—the "power PC theory"—is also statistical and probabilistic and works only in simple situations in which:

a. effects and candidate causes that are clearly defined and that can be represented in terms of probabilities, and

b. simple causes that influence the occurrence of an effect independently of background causes within a context. (p. 398)

These two conditions define a simple physical context in which the probabilities are high and the causal forces few. In such a situation, experiments can be carried out that allow us to test the necessity and sufficiency of the putative forces. Unfortunately, this approach does not hold for the multicausal situation present in virtually all psychological processes. In psychology, we do not have the luxury of controlling all of the variables or even sufficiently constraining our intuitions about possible causal relationships that are free of bias and expectations. Nor is it usually possible to conduct a conclusive experiment that establishes necessity and, even more important, unique sufficiency.

Statistical analyses involving correlations, therefore, can misdirect us to impute causation in situations in which it is but a phantom. For example, two variables may be highly correlated, but it may be impossible to determine if one is a cause of the other or, vice versa, the other is the cause of the one. Furthermore, no correlational statistic can distinguish between the causal influence of one event on another, on the one hand, and the joint effect of a third, often invisible and unrecognized, event on two highly correlated events. The problem of unrecognized influential third events is constant throughout the study of the kind of complex nonlinear systems typical of biological medicine and psychology.

Throughout psychological and medical science the possible influence of an unidentified third (or fourth or fifth) factor as the essential causal agent, as opposed to an identified agent, remains problematic. Extremely subtle placebo effects can invalidate the best-designed experiments and have to be rigorously controlled. Whether or not they can be adequately so controlled remains a consistent challenge, especially in statistical studies in which there is a vested interest in the outcome.

The most obvious examples are the long duration of the controversy over the health effects of cigarettes and the relationship between historical greenhouse gas concentrations and global temperatures. There are many other similar debates, particularly in terms of the health effects of foods or medicines or climate, that continue to plague modern societies. It is interesting to note that even when there are very high correlations between two variables (as in the CO^2 and temperature situation) and there are no other plausible or identified causal relationships, there is still ample room for dispute and controversy, especially if there are other unrelated vested interests at work.

Thus, even the best single-variable correlations are only suggestive. To go from the correlation to a convincing argument that the "cause" has been identified usually requires other kinds of empirical support. These supportive arguments include empirical laboratory studies in which the relationships can be manipulated and, in at least some cases, the development of sufficiently compelling formal models that lead deductively from one variable to another. This is what happened in the cigarette controversy and, hopefully, will happen before the impending climatic disaster is upon us.

However, it is not possible to unequivocally establish causal relationships from correlations alone. Some systems are recalcitrant to direct empirical examination, intrusive opening of a closed "black box," or adequate control. Just repeatedly replicating an experiment may leave controversy intact. Butz and Torrey (2006) optimistically suggest that there is hope that we can move from "complicated correlations to useful prediction" (p.1898) by linking such innovative strategies as "longitudinal surveys," "laboratory experimentation," "geographic information tools" (such as Global Positioning Systems), and "improved biological science tools." Notwithstanding the possibility that correlative statistics can make probabilistic predictions, there still remains an enormous logical gap between the ability of such measures on their own to identify causes. This is especially true in the context of multifactorial situations that characterize most kinds of psychology. Implicit in Butz and Torrey's (2006) concluding comment are some classic nonsequitors that Yule would have found unsatisfying:

> Those [innovations] discussed here have more potential than has been realized so far to attribute causality to observed relationships, to understand their nature, and thereby to improve the accuracy and usefulness of predictions. (p. 1899)

In my opinion, this statement conflates the fact that correlations can support "the accuracy and usefulness of predictions" with their ability to "attribute causality."

A further problem with correlative studies (as well as any other mode of analysis) is that complex, nonlinear systems are almost always characterized by feedback loops in which the later stages of the system influence earlier stages. Therefore, it is not always possible to distinguish the precursor causes from the resulting effect. What appears to be the locus of an upstream effect may actually be the source of a powerful feedback influence. Thus, the truly influential causal relationships are deeply hidden in the interactions of the various components of the system. In such a situation, identifying the components and the effective interacting driving forces may complicate the analytic problem to the point of mathematical intractability, no matter how much effort and resources are applied. This was shown to be the case for the visual nervous system by Hilgetag, O'Neil, and Young (1996; 2000). This near-hopelessness of unraveling causal sequences in complex systems exerts a powerful force on psychological experiments and theories by underdetermining the outcomes and thus opening the door to overly imaginative, if not fantastic, thinking and theorizing.

Although the study of the linear nature of simple physical and biological science has appropriately been driven by the historical admonitions of the Cartesian method, it is not easy to compartmentalize and separate a complex system into its parts. Most psychological experiments or more complicated physical systems are inherently nonlinear. The best we can do is to introduce a well-controlled stimulus (i.e., a presumptive cause) and to observe a well-measured response (i.e., an effect) and hope that any significant additional influences and potential biases have been adequately controlled. The worst we can do is to invent simplistic causal explanations of complex, nonlinear systems in which the stimulus is undefined and the response underdetermined. This creates a tension between many of the grand issues in psychology, ultimately to the ebb and flow of influence between the mentalist and behaviorist schools of thought.

Can causation be established in psychology and social sciences? The answer to this question, like that of the establishment of validity, is that, unlike gravitational theory, it is often more a momentary consensual judgment than the result of a rigorous proof. If we have to start someplace, we have to accept the basic sine qua non: the principle of linear causality. That is, there must be an orderly temporal sequence between a preceding cause and a subsequent effect. This basic principle is a necessary condition of physical science, but not a sufficient one; without demonstrable precedence, however, putative causality becomes virtually impossible to confirm. Unfortunately, linear causality is elusive in psychological experiments.

The other criterion for definitively establishing causality is a demonstration of the sufficiency of the putative causal agent to produce the effect by itself. Even then, it is not beyond the bounds of possibility that other causes may also be sufficient to produce the same effect. This very possibility inexorably leads to an open-ended search for these alternatives. Some might say that the current state of psychology, filled as it is with redundant microtheories, is the end-product of its essentially correlative, inductive nature and the absence of sufficient constraints to permit application of the axiomatic-deductive methodology that has proven so powerful in the physical sciences.

None of this is unique to our science. Psychologists share the classic problem with historians and other social scientists: How do any of us distinguish between multiple and equally plausible causal factors when all are equally well (or equally poorly) correlated with the observed effect? The essential part of this problem is that in the presence of only correlative information, the system is highly underdetermined, and insufficient constraints operate to limit hypotheses and theories. Thus, for example, the biologist and historian Turchin (2006), when considering the "causes" of the decline of the Roman Empire, argued that:

> Many things were going on in Rome, and an aspiring theorist who wants to launch a new theory has plenty of material to pick from. Was it Christianity? Or the decline of

> the "bourgeoisie," as another historian or Rome thought? Or imperial bureaucratization? Or lead poisoning? Latching on any particular correlation of the rich history of the Roman Empire, and basing on it a grand theory that explains its collapse is easy, but ultimately unsatisfying. Dozens, perhaps even hundreds, of such theories have been proposed over the centuries. This is not the way to do science. (p. 284)

If we change a few words here and there, Turchin's analysis[21] applies equally well to psychology and any of the other social sciences that depend on correlations. If we computed the number of psychotherapeutic "theories" that have been offered in past decades, there is little question that a number greater than "perhaps even hundreds" would result.

The absence of adequate constraints when statistical methods are used is much greater than with conventional mathematics. This is not a criticism of statistics; it is an inevitable result of complexity, a complexity shared by all of us in the human sciences. The forlorn hope that we can overcome this intrinsic complexity by mimicking the methods suitable for the simpler physical sciences is not likely to be fulfilled.

Ascribing causal relation in psychology, with its enormous complexity, feedback, and nonlinearity, as well as with its paradoxical cognitive responses, becomes extremely difficult, especially when our main strategic tool—correlative statistics—almost always results in underdetermined descriptions rather than proven inferences.

4.5 INTERIM SUMMARY

In this chapter, I have concentrated on the limitations of some of the mathematical and statistical methods of approaching psychological explanation and description. I have tried to show that statistics and human behavior share many common properties that are not evident in conventional mathematics. Therefore, the natural state of affairs is that statistics and psychology have a natural affinity that has drawn them together, just as conventional mathematics has tended to mirror physical phenomena.

Although statistics is clearly a subfield of mathematics and must follow all of the laws of arithmetic and logic, there are fundamental differences in its applications and those of conventional mathematics. Whereas statistics is inductive, conventional analysis is deductive. Whereas statistics is designed to handle highly variable data from complex, multicausal, nonlinear systems, conventional mathematics has evolved in response to the needs of relatively simple systems with much less variability in their responses. Even then, modern analysis has enormous difficulties in analyzing even relatively simple nonlinear systems and often accomplishes this task by applying simplifying constraints or approximation methods. When the going gets tough for ordinary mathematics, science often turns to statistical methods.

There are, furthermore, profound conceptual differences between the manners in which the two kinds of mathematics are applied. Statistics is deeply concerned with the problem at hand; in fact, it is only meaningful in the context of the data with which it is dealing. Conventional mathematics, although it is often specifically applied to a particular problem, can exist in an abstract world. Pure mathematics, unrelated to any real problem, is a perfectly respectable activity. Deductive mathematics is designed to deal with specific cases; specific cases are meaningless in statistics. One observation tells us nothing about the important properties of a set of data such as the probability of each category of response. Mathematics deals poorly with uncertainty; statistics dotes on it. Statistics is much more sensitive to the forces of bias (i.e., unaccounted-for influences) than is analytic mathematics.

Unfortunately, statistics' advantages are also the source of some of some of its greatest disadvantages. Statistics, by virtue of its most fundamental nature, loses the ability to say anything about individual responses or behavior. Its goal is to seek out overall patterns and, to the extent it succeeds in that regard, it fails to precisely predict individual cases. In other words, to the extent that statistics succeeds in meeting its goals, it tends to lose information in a massive way. This is the basis of the admonition "Information pooled is information lost."

There is a real question about the variability or even the randomness that is observed with statistical techniques. Is it fundamental or is it merely a reflection of the complexity of systems and the difficulty of identifying what are otherwise observable facts? We really don't know the answer to this question. However, from a practical point of view, it probably does not matter. "Hidden in complexity" or "fundamentally unobservable" meld into each other and produce equivalent practical constraints. Nevertheless, statistics gives us an initial approach to dealing with complex, nonlinear systems made even more complicated by ubiquitous feedback loops—systems like the human mind-brain.

All kinds of mathematics, including statistics, suffer from some common handicaps. One of the most overlooked of these handicaps is that the properties of the mathematics are sometimes attributed to the system being modeled. Thus, the system may have imposed on it fictitious properties that are not its own. One of the most misleading instances is that many of the attributed properties of a system may actually be due to the inevitable properties of the pooling processes typical in statistical analyses. Thus, logarithmic and power relationships are now believed by some researchers to be statistical fictions, rather than valid reflections of psychobiological processes. Power laws and 1/f functions, to cite two prominent examples, are now known to be the result of the pooling process itself; many other, quite different, functions characteristic of individual responses turn into power and log functions when data are pooled. In general, competent mathematicians believe the necessary tests to support one or another of these functions as "true" have not been carried out. Furthermore, even the best-fitting laws seem to be only coarse approximations to data.

Finally, the nature of our mental life makes it extremely difficult to determine causation. All agree, in the abstract, that correlation does not imply causation.

However, many psychologists apparently operate on the basis that it does. Unfortunately, even the most basic criterion for the attribution of causality—the phenomenal cause must precede the phenomenal effect—does not seem to hold for many cognitive processes. Given this situation, it may not be too extreme to assert that once beyond the coded and relatively simple sensory processes, it becomes not only difficult but impossible to identify the neurobiological or cognitive causes of a behavioral response! This may be the most fundamental reason that psychology and the other social sciences have turned to statistical analyses; statistics is fundamentally correlative and descriptive; only the foolhardy would attempt to use it to identify unique causes of our behavior.

Why are psychological responses so recalcitrant to identification of causes? The answer to this question is implicit in the complexity of the mind. Statistics helps us to describe that complexity and to summarize and condense it. However, it is not capable of enumerating, much less identifying, the causes. This limitation has to be added to those discussed in previous chapters in helping us understand why the nearby mind is so much less accessible than some of the remote entities dealt with by cosmological or quantum-level physics.

NOTES

[1]This chapter expands on some preliminary ideas expressed in Uttal (2003; 2005; 2007).

[2]Some scholars, such as Newell (1990) and Anderson (1993), have proposed what they call "general" theories. A close examination, however, suggests that these theories are either programming languages sometimes suitable to the representation of cognitive processes or are limited to a narrow range of psychological activities.

[3]The role of uncertainty in quantum physics remains problematic. Is it a true reflection of the nature of the world or only a limit on our knowledge?

[4]To clarify this point, a second-order differential equation of the form can be studied, manipulated, and solved independent of the specific meaning of the form

$$y = x + \frac{dx}{dy} + \frac{d^2x}{dy^2}$$

can be studied, manipulated, and solved independent of the specific meaning of the terms. However, each component maintains the same general meaning. For example, *dy/dx* always refers to a force that is proportional to the velocity or rate of change of *y* with respect to *x*. That rate of change may represent the movement of a spring, the rate of change of a population of fleas, or the decline in the concentration of a photochemical when acted on by light. In any case, the course of the analysis can be impeccable and enlightening without being embedded in a meaningful context. Statistics, on the other hand, is meaningless unless embedded in a specific applied context.

[5]There is a caveat to this generalization. According to the Pauli Exclusion Principle, no two electrons *in an atom* can be identical. That is, they cannot have the same quantum numbers. This idea has been generalized to explain why no two basic particles can be in the same place at the same time. From this point of view, no two electrons are identical, although defined by the same parameters. Psychological states do not enjoy this same level of simplicity.

[6]Those interested in a complete history of mathematics are directed to the very readable book by Boyer (1989) or the encyclopedic review edited by Gellert, Kustner, Hellwich, and Kastner (1977).

[7]This definition would exclude statistics and is much too restrictive.

[8]The personal equation work is especially significant for psychology. It was the first instance in which distributions of human responses (in particular, reaction times) were appreciated as being necessary to solve an encountered problem: variability in measuring astronomical transit times. This particular development, according to Mollon and Perkins (1996), led eventually to the establishment of the first laboratory of experimental psychology, if not the science itself. Does this suggest that Bessel, rather than Wundt, should be credited with the creation of scientific psychology?

[9]Stochastic is another general adjectival term describing any process that is governed by the laws of probability and randomness.

[10]The repeated use of the word "data" here is intentional. If there is any single property of statistics that distinguishes it from other forms of the usual kinds of mathematics, it is that it is meaningful only in the context of empirical findings (i.e., data). I expand later on this essential distinction between conventional mathematics and its spin-off: statistics.

[11]Traditionally, random sequences were generated by selecting random numbers from tables in mathematical handbooks. In many modern experimental laboratories, computerized random number generators operate on the basis of an algorithmic program to produce a pseudorandom number series—i.e., a series of numbers that are not likely to repeat in the short run. It is well known, however, that computerized random number generators starting from the same seed number always produce the same sequence of "random" numbers in the long run, and thus may introduce bias into the results of an experiment. Selecting a new seed for each procedural step or trial is, therefore, always a desirable second order of defense against subtle sequence effects.

[12]There are, of course, situations in physics in which uncertainty or randomness does play a part. Quantum mechanics and gas dynamics are two obvious examples. The confirmation of an infrequently observed event (such as that necessary to prove the existence of a neutrino) is another. The difference between physics and psychology that I paint here is of degree and extent, not of the nature of their common reality.

[13]This debate is not, of course, limited to the social sciences. It was, and is, the basis of the continuing debate between the probability theorists of quantum me-

chanics and the determinists, such as Albert Einstein, who argued that "God would not play dice with the Universe."

[14]It is important to appreciate that the 1/f function, fractals, and power functions are all closely related. What is said about the origins of the 1/f function (that it is the result of the dynamical processes of interaction between entities rather than the properties of the entities themselves) also applies to power laws.

[15]The analogy may be drawn between the infinity of possible wavelength combinations that can produce the same average value on the CIE chromaticity diagram and thus the same color experience (isomeric colors). The actual information that led to a particular color experience is lost just as the original data that collectively led to some measure of central tendency is forever lost when its effects are combined.

[16]See also the discussion on page 124 in which the power law and 1/f functions are related.

[17]In doing so, Narens implicitly joins those of us who argue that behavior is not transparent to its cognitive underpinnings.

[18]This is another example of how even good fits between psychological process and mathematical equations may be misleading. Roberts and Pashler's (2000) admonition is particularly germane in this context.

[19]In fact, the exponent for electrical stimulation was much greater than 1.0. For the somatosensory system it was about 3.0. Obviously, such an expansion of the response with increasing stimulus intensity could not be maintained and is not realistic. Exponential growth cannot continue indefinitely without some kind of an explosion. However, this expansion does emphasize the dramatic effect of bypassing the compressive function of the receptors.

[20]Tufte (1983) also warned us about the problems that can be introduced into scientific theorizing when the conveniences of graphic usage are confused with physical or psychobiological processes.

[21] Despite these insightful words, Turchin (2006) does spend a major portion of his book defending one particular explanation of both the rise and the decline of the Roman Empire: the fact that it had boundaries with different ethnic peoples with different cultures. Turchin at least appreciated that alternative theories are possible in underdetermined situations.

5

General Conclusion: How Cognitive Inaccessibility Denies Inferential Power and Influences the Great Debates in Psychology

5.1 INTRODUCTION

This book has reviewed a number of the reasons why physical science can draw inferences about inaccessible physical processes, but psychology cannot do the same for its inaccessible mental processes. The reasons are manifold. To briefly summarize:

- The properties of space and time are profoundly different for physical and mental processes, respectively. Physical time and space are ordinal, monotonic, continuous, usually isotropic, and homogeneous. Psychological time and space exhibit paradoxical, disorderly, anisotropic, and elastic properties in a wide variety of phenomena that seem to conflict with those of physical time and space.
- There is nothing in the psychological domain that is the equivalent of the Cosmological Principle in physics. This key assumption asserts that the laws of physical nature are the same wherever in the universe we may look. It is this fundamental concept that makes it possible for physicists to break

through the practical barriers of time and space in order to draw inferences about distant or microscopic entities that are not directly accessible from their observations. Its absence (or the absence of any other comparable linking principle) is the reason for our inability to infer the nature of inaccessible mental mechanisms and processes from observations of behavior.

- The use of numbers, the basic arithmetic properties, and quantifiability in general, all differ in the two domains. Physical dimensions are characterized by non-arbitrary zeros, ratio scales, and adherence to the laws of arithmetic. These basic properties, which are required for quantification according to such scholars as Coombs (1950) and Michell (1999), are often absent in mental phenomena. Psychological scales "typically" (as opposed to "infrequently" in physics) have arbitrary zeros and non-ratio properties. In our cognitive worlds, time and space can be distorted by the observer in ways that are in opposition to the most fundamental physical principles. Furthermore, the basic laws of arithmetic are not always followed by mental phenomena. Most serious of all, many psychological phenomena exhibit paradoxical time inversions in which the cause appears after the effect. This means that one of the foundation principles on which physics is based—linear causality—is often violated.
- Robust science depends on robust metrics. That is, the quantifiability of a dimension requires that the underlying metric (the "geometric function that describes the distances between pairs of points in a space") be regular, at least to the extent that we know how it may systematically vary. Unfortunately, the metrics of psychology are often obscure and undefined. This raises serious questions about the nature of what we mean by measurability when dealing with mental phenomena. An argument can be made with some vigor that, since they are not quantifiable, mental parameters are not measurable in the same sense of physical dimensions.
- Mental phenomena are dominated by processes that are best described by probabilistic or stochastic terminology whereas ordinary non-quantum physics generally is best described by a deterministic vocabulary. One result is that psychological activities are best represented by statistical models and physical ones by conventional analytic models. Statistics is mainly an inductive method in which the context gives meaning. Conventional mathematics can survive in the abstract, with context considered only secondarily. Statistics is a correlational approach in which not only causation may be very elusive, but the involved causal parameters may be difficult to identify. Analytic mathematics deals much better with situations in which the result is the outcome of a few initially well-defined parameters, forces, and causes. The reason that psychology and statistics have drawn together is because they share common properties and features. The reason that physics and analytic mathematics have such an affinity for each other is exactly the same.

- Both conventional mathematics and statistics are beset by some little appreciated handicaps. All mathematical models are neutral in principle with regard to the inner workings of the system being modeled. Mathematics can inject fictional properties into the analysis either from super-powerful analytic techniques or inevitable outcomes of processes such as data pooling. The properties of individual responses may not be maintained when the data are pooled. For example, exponential functions may add together to produce power functions. Statistical studies are particularly, but not uniquely, susceptible to these disadvantages.
- The differences between physics and psychology are in part explained by differences in the complexity of the causal forces imposed in each case. In psychology we rarely are able to measure the isolated impact of an imposed force, much less identify causal forces. The issue of how we deal with inaccessibility ultimately may be resolved differently for each domain of science. Nevertheless, both physics and psychology deal with entities that are the outcome of natural causes and events; there is no need to invoke supernatural forces in dealing with the mind any more than there is when dealing with the trajectory of a missile. Unfortunately, scientific psychology and its offspring, cognitive neuroscience, have much more difficult challenges than does physics because of the complexity of our neural and mental processes. It has responded to its special problems by adopting a statistical mode of analysis because the much simpler road to understanding physics—axiomatic deduction—is not usually available to it. The best evidence for this is that there have been no axiomatic-deductive theories of mental function that have stood the tests of time.
- At the very bottom of this logical chain is the undeniable fact that the empirical data of psychological science discussed in Chapters 2 and 3 repeatedly demonstrate that our cognitive processes are not veridical with the parameters and dimensions of the physical world. We cannot anticipate what the response to a physical stimulus will be, nor can we predict individual behavior, both accomplishments being among the crown jewels of physics and all other normal sciences.

These conclusions and judgments would only be of interest to those interested in arcane and esoteric aspects of speculative philosophy, except that they provide the conceptual foundations for understanding the current state of scientific psychology. The purpose of this concluding chapter is to explore how these seemingly abstract conclusions affect the day-to-day activities of both empirical and theoretical psychologists. In particular, I now examine how these properties influence our ability to answer some of the great questions that have perplexed psychologists and their predecessors for many millennia.

In searching for the key questions and issues, I turned to two of my previous books to identify those deserving of special attention. In Uttal (2001, pp. 221–225)

I appended to the book a list of 95 "great questions of scientific psychology." These questions were categorized essentially according to how they pertained to some of the tasks we have set for our science. These categories included:

- Theories
- Mentalism and Behaviorism
- Mind-Brain Relationships
- Mind
- Sensation and Perception
- Learning
- Cognition
- Miscellaneous

In Chapter 4 of Uttal (2007), I refined the list of questions and rather than organizing them about the subject matters of psychology, suggested a minitaxonomy based on the kind of principles on which a scientific psychology should be, and is, founded. This classification scheme included the following categories:

- Ontological Principles
- Epistemological Principles
- Statistical and Measurement Principles
- Methodological Principles
- Pragmatic Principles
- Empirical Laws

In the following paragraphs of this final chapter, my goal is to use these lists to explain how the limits on inference due to mental inaccessibility impact some of the most persistent and significant questions of psychological science.

I have used two criteria to cull these two lists of questions and issues down to a manageable size. First, I am going to finesse those that I designated as ontological issues. This group includes matters based on a priori assumptions that are not amenable to evidentiary, logical, or mathematical argumentation. No matter how massive the evidence or how complete its absence, nobody is likely to be driven from their deeply held beliefs concerning these questions by any argumentation. Mathematicians and logicians, physicists and biologists, can argue until they are blue in their academic faces about the absence of evidence supporting life after death or the existence of a supreme being without making a dent in the belief structures of the ardently religious or dualist. Similarly, committed materialists are unlikely to change their belief structures because of some allusion to spiritual values.[1] Even a "divine revelation" would be attributed to some brain disorder by the devoted material monist.

Some of the most important of these great ontological debates (not discussed further here) include:

1. Material events provide a complete explanation of mental events, versus mental processes represent a separate level of reality that can exist without the material substrate.
2. Mind terminates at the moment of physical brain death, versus our "souls" continue to live on after bodily death.
3. Mind is instantiated by brain functions, versus mind is extended throughout the rest of the nervous system and perhaps beyond that to the rest of the body or even to the environment.
4. Cognitive behavior is determined by single causes, versus cognitive behavior is determined by multiple causes.
5. Mental life is fundamentally stochastic in the same sense as quantum mechanics ("God *does* play dice with the universe"), versus mental life is fundamentally deterministic ("God *does not* play dice with the universe").
6. Randomness is just an expedient until we can identify all of the hidden variables, versus randomness is an expression of a fundamental uncertainty.
7. Mind and free will are real, versus mind and free will are only illusions or phantoms.
8. All of behavior is composed of variations of a few basic themes (e.g., reflexes and classical and instrumental conditioning) subject to scientific examination, versus all behavior is the result of immensely complex neural and cognitive interactions beyond our ken.

Unfortunately, some epistemological arguments are equally difficult to distinguish from ontological beliefs about the natural and supernatural world simply because we do not have enough evidence to support one side of an argument or the other. For example, it is generally assumed that the popular concept of the "mind" (whatever it is and however it is defined) is the outcome or product of the huge neural network of the brain. This network is characterized by the individual idiosyncratic interaction of many billions of individual neurons. This hypothesis seems extremely plausible; nevertheless, it is likely that it, too, can never be "proven." There are no mathematical or computer models that can test it; no neurophysiological evidence is available to robustly confirm it; and no correlation is sufficient to prove both its necessity and its sufficiency. All we have are plausible ideas and suggestive, but not definitive, observations.

This state of affairs means that there is only a minimal likelihood that an incontrovertible ontological theory of the origins of the mind from brain activity will ever be forthcoming. Indeed, I have argued (Uttal, 2005) that the mind-brain problem may never be solved because of the complexity of the brain mechanisms (in the form of huge numbers of neurons and even larger numbers of interconnections) that give rise to sentience. Of course, it is always difficult to say that some great scientific advance in the future may not provide a better understanding and even a definitive answer, but the mathematics of the situation suggests that the problem, like many of the ontological issues, is intractable.

The second criterion that I use to identify the issues to be considered here is my personal judgment of their importance. That is, I select those issues, problems, and disputes in psychology that are of historical and global consequence. There are many methodological and even first-order theoretical questions that I do not choose to deal with here. Among these "quasi-trivial" questions are those implicit in the specific hypotheses that we seek to answer in our day-to-day laboratory experiments. Instead, I concentrate on those that I believe are of transcendent importance, not to a single experiment, but to the entire science. Among them are some of the classic conundrums that have challenged thinking in this field for generations.

5.2 CAN MENTAL PROCESSES AND MECHANISMS BE INFERRED FROM BEHAVIORAL OBSERVATIONS?

Of course, the most basic and repeatedly asked (implicitly if not explicitly) question throughout this entire book deals with the plausibility of inferring mental processes and mechanisms from the observations we make of behavior. For a host of reasons already described in previous chapters of this book, I believe that there are profound differences in the quality of the inaccessibility faced by psychology and physics, respectively. Whereas physics can assume that the laws "here" are the same as laws "there" (an assumption that is either synonymous with or a direct derivative of the Cosmological Principle underlying the success of physical science), the mental and cognitive processes of which we have only the barest understanding regularly distort, obscure, and conflict with the laws of physical time, space, and number. Not only are the laws of physics seemingly violated during mental activity, but there are also other barriers to reductive analysis, such as the one-to-many problem: the *fact* that there are many (perhaps innumerably many) explanations of any observation when the system under study is closed or otherwise inaccessible. This has a very powerful implication—namely, that all observations are indeterminate with regard to underlying structure. This barrier to explanation is particularly profound in psychology, where the only clue to otherwise inaccessible mental activity is behavior.

What this means is that it is not, in general, possible to infer the nature of the specific cognitive, mental, psychologica l, or physiological processes that account for our behavior from that behavior. The implications of this conclusion, if it is correct, are widespread. For example, it means that much of the effort to explain (as opposed to describe) perfectly sound behavioral observations is wasted. We can never, according to this point of view, develop an exclusive reductive explanation of behavior because the necessary conditions (i.e., the data) for deriving the nature of those mechanisms do not exist. That is, observed behavior, no matter how precisely measured, underdetermines the exact nature of whatever underlying mechanisms account for it. If this were a mathematical problem, we would say that a unique solution is not obtainable since the problem is "ill posed." A return to the basic tenets of a more or less radical behaviorism is, therefore, overdue. Although I appreciate that there is no "killer argument" that this conclusion is correct, it seems logically sound and has not yet been empirically falsified.

One implication of this analysis is that psychology made a major strategic error in changing to mentalist, cognitive, reductionist approaches from classic behaviorism. In conjuring up a host of hypothetical constructs and then reifying them, it has seriously misled scientific psychology and reduced the prestige of our science in the overall scientific community.

However much acceptance this suggestion may have, there are some caveats that must be immediately made to make my position clear and to avoid any suggestion of nihilism, pessimism, or, even worse, supernaturalism.

1. By no means is anything I am suggesting here contrary to the basic ontological assumption that mental functions are processes of a physical entity: the nervous system. The fact that we are incapable of making the leap from the behavioral to the mental or the neural does not mean that there is anything other than a practical boundary preventing cognitive- or neuro-reductionism. The difficulty (or even the impossibility) of identifying the way in which the nervous system produces the mind cannot be used as a crutch for inserting supernatural concepts into the discussion.
2. Nor am I suggesting that the empirical data base collected by psychologists over the last century is anything less than important or useful. Those data consist of observations and descriptions of behavior that stand on their own feet as major accomplishments of our science. It is what we infer (in the form of theories or hypotheses) from these findings that is the crux of the critique I present here.
3. However robust behavioral observations may be, the extrapolations from data to hypothetical constructs and imaginative explanations are rarely justified. Many, if not most, of the theories and reductive explanations (by which I mean those that try to explain the observations in terms of neuro-physiological or cognitive components) are unsupportable whimsies stimulated by the underdetermined nature of behavior observations. Occasionally, one explanation may be shown to be correct in some narrow context (such as peripheral sensory communication). However, such a "hit" may be entirely fortuitous; we have no way of distinguishing the valid from the invalid in a chronically underdetermined system. There is insufficient information and there are insufficient logically robust bridges to make the leap from behavior to mind. On the other side of the coin, there are too many neurons to bridge the gap between behavior and brain mechanisms and processes.

If this argument is correct, then it has serious implications for other issues that have long concerned biologists, philosophers and psychologists, issues to which I now turn.

5.3 CAN A TAXONOMY OF COGNITIVE PROCESSES BE ORGANIZED?

If the development of a comprehensive theory of even a modest selection of psychological phenomena seems elusive, perhaps we might next ask: Is it possible to at least organize the field into an ordered classification scheme? At the most primitive level, of course, we might just name behavioral observations (a typology) and describe the features of each type. Typologies do not seek out any of the relations between the various types; systematic organization and causal interactions are not sought or achieved.

If, however, systematic relations between the various types are discovered, then we can organize the "types" into the next higher level of organization: a taxonomy. During the early history of taxonomic classification, scholars such as Carolus Linnaeus (1707-1778) examined what was then just a collection of types and observed that there were some observable relationships that permitted them to be organized on an ad hoc basis. Their systems were not analytic; they made no attempt to explain the source of the relationships, but they did find some dimensions of similarity along which to arrange their specimens.

In recent years, however, the development of taxonomies has evolved into a rigorous science known as Cladistics. Currently, cladistic taxonomies are produced by the application of a formal computation approach, originally proposed by Hennig (1966). The application of cladistic programs is not simple or universal. It works only under certain conditions. These conditions or assumptions are usually expressed as follows:[2]

1. The entities or types (e.g., organisms) at each level must be traceable back to a common source.
2. Descendent types appear strictly by bifurcation.
3. There is a progressive series of changes in the characteristics of types from level to level. In biology this may occur over time; in psychology this condition may simply mean that there is an organized tree of phenomena in which the characteristics change from level to level.

The first of these three conditions for cladistic analysis demands that there be some relationship between the various levels of the taxonomic tree. Thus, it is required that phenomena not be independent of each other. For a science of anything to be organized into a taxonomy, there must be relationships of causation, shape, or some other dimension that permit the ordering of the phenomena into a meaningful schema. This is an absolute necessity; if we had no organizing dimension or principle, then it is a truism that nothing can be done beyond the assemblage of a typology.

The second condition is mainly a practical and simplifying one. It is inserted because of a computational problem. If the division of the ancestral clads (categories) is not made into two (as opposed to three or more) subtypes, then the computational problem would explode and become intractable. This condition is also influenced by the fact that in biology, types seem to have appeared as one of two possible descendents from a single ancestral organism.

Finally, the third condition is the heart of cladistics. Descendent types must be different from the ancestor, and these changes must progress in an orderly way through the taxonomic tree. A change at one level must be related to, and possibly preserved in, the characteristics of higher levels.

The question now arising is: Do psychological phenomena meet these conditions sufficiently well to maintain the hope of an orderly taxonomy? The answer to this question is equivocal and depends on what field of psychology is of concern. Organized arrangements of some limited fields are possible. I tried my hand at organizing the field of vision in a book entitled "A Taxonomy of Visual Processes" (Uttal, 1981). However, as one moves away from the peripheral sensory communication processes and deals with high-level cognitive functions, the development of taxonomies becomes much more of a challenge, if not an impossibility.

There is a considerable amount of suggestive evidence that psychology in general has so far failed to organize itself into a well-ordered taxonomy, as opposed to an arbitrary collection of unrelated types. The absence of any trend toward consolidating data into a pyramiding theory is one such piece of evidence. Another is the inherent and ubiquitous disorganization of psychology textbooks at all levels (in which chapters do not lead progressively from one idea to the next). The endless proliferation of hypothetical constructs, microtheories, and irreconcilable controversies suggests that the relationships between the causes of various phenomena are actually invisible.

A major implication of the absence of an organized taxonomic scheme for psychological phenomena is that the quests for modular cognitive components and for localized brain regions encoding these components are ill advised. If we do not yet have anything more than a typology of cognitive components, it seems likely that their precise definitions have not yet been achieved. Rather, many hypothesized cognitive components, mechanisms, and processes are little more than names of experimental results or unsupportable inferences from behavioral observations. The effort to locate such a rabble of phantoms in particular regions of the brain seems quixotic, at best.

The most obvious and complete failure to produce a satisfactory taxonomy of mental entities probably is the Diagnostic and Statistical Manual of Mental Disorders (DSMV-IV) (Anonymous, 2000), an effort to organize the diagnostic relationships among mental illnesses. A close analysis of this document (see, for example, Spiegel, 2005) makes it clear that it is not anything more than an expanded typology that was collected on the basis of committee discussions, rather than the specific identification of relationships in empirical studies.

Thus, a taxonomy of modular or semi-discrete psychological entities is not at hand. Although there is no way to exclude the possibility that such a system may be created in the future, it clearly is not a major goal of psychological research today; few psychologists are at work trying to organize our huge data base into some preliminary coherent classification system. This lack of commitment may be due to our passionate love affair ("physicophilia") with experimentation. Conversely, it may reflect the fundamental fact that the system simply does not display the relational order necessary for a cladistic analysis. Since a taxonomy seems to be a necessary precursor for a theory, the absence of such a cladistic scheme for psychology is another argument that psychology is not ready for a comprehensive theory.

5.4 CAN ANIMALS THINK?

Of all the hoary old questions of psychology, perhaps none has raised such continuous popular interest as the questions of animal consciousness. To put it in a slightly more formal context: Can we distinguish between a sentient, logical, conscious entity and a mindless, unconscious automaton?

Despite a considerable amount of wishful thinking and a considerable amount of research on gorillas, dolphins, parrots, and other fascinating and interesting creatures, there is no conclusive evidence that animals are aware or exhibit conscious decision making. On the other hand, there is no evidence that they do *not* experience at least a reduced version of the rich perceptual and emotional sentience reported by humans.

The problem is the familiar one: Behavior does not provide the bridge to the mind that is assumed by students of animal cognition. Exactly the same behavior could be produced by a mindless automaton and a conscious creature. Furthermore, there is no basis for arguing that animal minds follow the same laws as the human mind in a way that permits valid inferences from the analogies of behavior. Given that we cannot infer the nature of the mind from behavior, the problem will always be of great, but irresolvable, interest.

Indeed, the problem is broader than just being limited to the issue of the animal "mind." This classic problem has now reappeared in a modern context: Can a computer be conscious? The problem in this case is the same as the animal thought issue. How could you tell? What clues are there to sentience that distinguish it from automatic behavior? The answer is that there is probably no way to tell the difference, and the debate will rage on far into the future, when computers are behaving in manners comparable to that of the human. Despite a popular belief, there are no Turing tests that can distinguish between an aware entity and a machine.

This is not the full extent of the issue, however. If we cannot tell if an animal or a computer can think, why are we so assured that we humans think? "Cogito ergo sum" ("I think, therefore I am"), Descartes' well-remembered admonition, is of little help. It assumes individual reality on the basis of a self-awareness, but that self-awareness does not meet one of the main criteria of scientific argumentation: public availability. Indeed, it might be more useful in the present context if reversed into "I am, therefore I think." That is, because of my awareness of my existence, I am assuming that I think!

The problem is that there is only one argument, one piece of data that suggests that humans are actually sentient. That is our own personal experience of our awareness. Personal self-awareness, however, is a private matter that gains even modest credence only because of the limited ability we have to articulate our ideas and exhibit intent. Each of us has first-hand evidence of our own consciousness, but there is no comparable evidence of awareness in any of my fellow humans. The best I can do is to draw an analogy to other minds from my awareness of my own. In point of empirical research, the question of human consciousness faces the same problem that computer and animal consciousness do. There is no way to distinguish between automatic behavior and active and sentient cognition.[3]

There is an additional practical consideration. We have a strong need to assume that others are sentient. It is necessary to use a mentalist vocabulary and to use words such as "me," "you," and "think" to avoid making the entire human enterprise meaningless.[4] Perhaps Descartes would have been even more correct if he had acknowledged this social need and said, "We are, therefore we think!" This revision suggests an intrinsic need for interaction among individuals. Unfortunately, any effort such as this leads us off into a verbal muddle from which there may be no escape.

5.5 IS A UNIVERSAL THEORY OF PSYCHOLOGY POSSIBLE?

Physics, the role model if not the mistress of psychology,[5] has advanced hand in hand with mathematical theories. Newton's (1687) great *Principia* set a tone for modern physical theory that persists to this day. Starting from a few general principles and three sharply defined axioms (his three laws[6]), Newton was able to prove a variety of theorems about mechanical systems and thus describe with great precision how certain phenomena occurred. His explanations included such powerful concepts as the inverse square law and why the orbits of the planets are elliptical rather than circular. His enormous success was due to three factors: (a) the application of a well-established deductive set of mathematical laws; (b) the relative simplicity of the system that he was seeking to describe; and (c) the quantifiability of the space, time, and force dimensions with which he dealt. In addition to the simplicity introduced by his three axiomatic laws, there was only one major force—gravity—that he needed to incorporate into his theory. It was this single form of attractive force and its properties that permitted one of the most amazing accomplishments of human intellectual history. In short, deductibility and simplicity were the keystones of Newton's contribution and those of many comparable physical theories to follow.

It would be wonderful if psychology were able to emulate this approach to the extent dictated by our physicophillic passions. The desire to build a model of the psychological world comparable to the one Newton and his predecessors and successors built for the physical universe is a recurrent, if unachieved, theme of psychological research.

Alas, the few instances in which such an attempt was actually made have eventually imploded. The most famous of all efforts to build a Newtonian-type system for even a restricted field such as learning was Hull's (1943; 1952) "mathematico-deductive" mimicking of Newton's Principia. Unfortunately, Hull's energetic effort failed to come to the same kind of influential conclusion that Newton's did. Since the mathematics was not only available, but even improved since Newton's time, and Hull was in all likelihood intellectually capable, his failure can only be attributed to the nature of the system he was studying. Koch (1954) enunciated what he believed were the main causes for Hull's grand failure of accomplishment, if not of ambition:

1. Secure anchorage [to observable and measurable conditions and events] either in a quantitative or qualitative sense, [did] not hold in a single case for the relations of systematic independent and dependent variables to their intended range of reductive symptoms.
2. No given intervening variable is securely and unequivocally anchored to its relevant systematic independent and/or dependent, variables either quantitatively or qualitatively.
3. No given intervening variable is related to any other intervening variable in the chain with sufficient determinacy to permit quantitative passage from one to the other, nor are certain of the variables, and the relations connecting them, defined with sufficient precision to permit "qualitative" passage. (p. 160)

Koch's first and second points argued that the conditions for quantifiability are not adequately met in Hull's theory and by implication in psychology more broadly considered. His third point essentially suggests the laws of deductive logic and mathematics do not apply to Hull's theory. All three collectively point to the alarming (for psychologists) conclusion that psychology is too complex (or for some other poorly understood reason) and incapable of applying the axiomatic-deductive system that has served physics so well.

We are thus led to the conclusion that the hope of a universal or even general axiomatic-deductive theory of psychology in the spirit of Newton's work is not fulfillable. The classic tradition of a science evolving from a large collection of independent observations to an encompassing small set of general principles and laws facilitated by a deductive mathematical methodology seems beyond our grasp at the present time. Certainly, there is no evidence of such a pyramiding in modern scientific psychology. The laws of psychology are imprecise and at best approximations. Our theories are, to an almost total extent, underdetermined by our observations. The psychological scene is increasingly cluttered with more and more hypothetical constructs that are not "securely anchored to observable and measurable conditions and events either in a quantitative or qualitative sense."

Again, it is important to appreciate that this is probably not the result of an inadequate system of analysis or of a paucity of data. Instead, it is more likely the result of the nature of mental activity. Hull's effort failed, and all others that aspire to be like it also will inevitably fail because psychology is what it is, not because it is a young or primitive endeavor yet to achieve its maturity.

One of the main pieces of evidence to support this disappointing conclusion is that the methods of statistical analysis fit the needs and properties of the psychological sciences better than conventional mathematics. Statistics, as we know it today, is not a deductive approach to understanding; it is an inductive approach that cannot identify constituent forces or disentangle multiple causes. The best we can do is to show that some stimuli are more effective than others in producing some response. How the stimuli are transformed and how they exert their influence, however, remains uncomprehended and most likely incomprehensible.

What statistics does best is to describe the typical behavior of complex systems in which multiple parameters, dimensions, and forces collectively and interactively determine the outcome. Stochastic variability may or may not be the underlying reality. This, as mentioned earlier, is an ontological and probably irresolvable issue. Whatever accounts for the quasi-random nature of psychological responses, they are the result of the extreme multidimensional and multifactorial complexity of the human nervous and cognitive systems. In sum, our mind-brain is a system of such complexity that it prohibits the development of an axiomatic-deductive "Principia Psychologica."

5.6 CAN PSYCHOLOGICAL PHENOMENA BE MEASURED?

There is implicit in the previous question an issue that may be of an even more fundamental nature than: Can we build a theory? That more basic issue is: Can psychological processes be measured in the quantified sense that physical dimensions can? Given what the conditions are for measurement, as spelled out by Michell (1999) and Coombs (1950), there is a reasonable argument that can be made that the offensive and disheartening (to psychologists) conclusion that mental processes cannot be measured in the same sense as a physical phenomenon is valid.

Before I consider this issue, it must be made clear once again that I am not referring here to observed behavior. There is no question in this thesis that all of the conditions of quantifiability for measuring behavior are met. Ratio scales, non-arbitrary zeros, and arithmeticity of the dimensions of behavior are all present. However, whereas both the quantifiability and measurability of behavior are unquestioned, the same cannot be said for mental processes and mechanisms being carried out in an environment in which the essential conditions for measurement are not met.

The very quantifiability of our inferences from behavior to mind is, therefore, challenged. Without the minimal conditions for quantifiability, measurement of mental events becomes as evasive as the experiences of the subjective illusions and paradoxes of time and space themselves.

This lack of quantifiability and measurability is why it has proven so difficult to develop scales of human mentation. Even in the simplest cases of sensory magnitudes, suggested units of measurement such as "brils" or "sones" have quickly been discarded as being meaningless when it became clear that they lacked equal intervals, ratios, aritmeticity, and non-arbitrary zeroes and metrics. All that can be done is to show that there is some kind of a correlation between some attribute of the response and some attribute of a stimulus. This kind of correlation is not the same thing as a quantitative *measurement* of an experience.

Without susceptibility to the laws of arithmetic, it is further clear that no mathematical models can be developed that can validly represent mental dimensions. The best we can do is to apply the methods of statistics to the task. Probability of occurrence is an artifice we use to overcome this enormous handicap of limited measurability of mental phenomena.

5.7 CAN ALTERNATIVE THEORIES OF PSYCHOLOGICAL PROCESSES AND MECHANISMS BE DISTINGUISHED?

The question must now be asked: if quantitative measurements cannot be validly made of mental phenomena, is it possible to distinguish between alternative theories or explanations of these phenomena? This is a critical issue because a positive answer to this question should be the foundation of any kind of science, not only psychology. It is widely assumed that psychologists can carry out an experiment in a manner that permits us to discriminate between alternative inferred explanations of some behavior. However, a close inspection of a broad range of literature shows that this rarely happens. Rather, the best we do is watch interpretations shift as the intuitive plausibility of one theory or another increases or decreases, a process not too different from arguing by anecdote.

I argue now that efforts to distinguish between different theories cannot be definitive because, at best, we are arguing over inductive inferences (i.e., hypothetical constructs) rather than deductive derivations from specific axioms and tried and true logical and arithmetic laws. For example, when we observe that people exhibit change blindness (a behavioral observation), there is no way that we can determine if the "gorilla" (see page 101) was really never seen or simply forgotten because of the competing demands on the observer's attentional resources. Psychology is filled with similar plausible but indistinguishable explanations of many other mental activities. As much as we might try to resolve a debate between two equally plausible, and equally underdetermined, psychological theories, it is not likely that the one will ever be sufficiently distinct to be rejected in favor of the other.

The main reason for the indistinguishability of alternative mental theories, as I have noted earlier, is that behavioral observations are underdetermined. All closed systems are subject to the "one (behavioral observation)-to-many (possible mental mechanisms)" difficulty. That is, there are an innumerable number of possible and plausible explanatory instantiations of any observable behavior or cognitive phenomena. The data from a psychological experiment do not contain the information necessary to produce definitive answers to questions of explanatory mechanisms and processes.

This fact is generally known in engineering and automata theory (Moore, 1956) but almost universally ignored in psychological circles. As already discussed, underdetermination leads directly to indistinguishability. In such a situation, we have to turn to secondary criteria such as simplicity or elegance, properties that, however pleasing aesthetically, are not robust criteria for choosing between alternative explanations.

Even in those cases in which it appears that we have "opened the closed system," it remains debatable whether alternative theories can be distinguished. Coltheart (2006a; 2006b) was especially critical of the ability of the newest wave of neuroimaging to distinguish between alternative cognitive theories. He pointed out that there were a number of logical flaws in many of the studies that used PET and fMRI scans in an effort to choose the "correct" theory. These included articles describing research in such problem areas as:

- Recognition memory
- Unexpected memory testing
- Inattentional blindness
- Facial identification
- Working memory
- The representation of other people

Coltheart analyzed the illogic leading to the choice of one explanation or another in these classic problem areas. Some of the flaws he noted are:

- The lack of specific predictions from either theory.
- Supporting results for each of the alternative theories.
- No robust support for either of the alternative theories.
- Lack of specificity of the imaged response.
- Irrelevant observations. For example, neural theories that do not speak to the cognitive theories or cognitive theories that do not speak to neural theories. (This is the bridging problem; neural and cognitive theories are compared by

means of superficial analogies without adequately establishing the homological links.)

- Neural modularity theories do not map directly onto cognitive modularity theories. Thus, even if we could demonstrate neural modules, it would not distinguish between two alternative cognitive module theories.
- No or inadequate consideration of any alternative theory.
- Conflated tasks.

What remains is a kind of wistful hope (as expressed by Schutter, de Haan, and van Honk (2006) that this new technique will eventually permit us to bridge the gap between the mind and the brain.[7]

If there are major problems in distinguishing between theories even in this area of brain imaging with its much greater tangibility, what hope is there for the much less constrained situation in which purely inferred cognitive explanations are tested against each other? It does not take too deep an inquiry to discover that many experimentalists already appreciate that their designs are unable to definitively resolve inter-theoretical disputes. Rather, the phrase "[observation x] is not inconsistent with [theory y]" (or some paraphrasing of it) is very common in the psychological literature. There are two possibilities for this ubiquitous expression: (a) Scientists should be appropriately cautious; or (b) It is a realistic expression of the fact that no behavioral data can ever confirm or reject a particular cognitive theory. If the latter, we must accept the unhappy fact that cognitive theories or explanations are fragile and unverifiable insights at best. This is what partially explains the unending number of unfulfilled quests to explain how our minds work.

5.8 OTHER IRRESOLVABLE ISSUES

Some of the most frequently asked questions of psychological science suffer under a cloud of irresolvable uncertainty. This does not mean that we have not made some progress in apportioning the variance to which the various factors may be attributed in some cases. But it does mean that identification of the specific forces involved in explaining our behavior remain, for the most part, elusive. Among the most notable of these major, but unresolved, issues are:

- Nature versus nurture
- Serial versus parallel processing
- Holistic versus feature processing of stimuli such as faces.
- Learning by means of experience versus learning by rational explication.
- Are ethics and morality intrinsic or social constructions?
- Are gender or racial or age differences responsible for behavioral differences?
- Reward versus punishment

Although psychologists will probably argue for generations to come about these and related disputes in our science, many of them are likely never to be resolved in the detail that physical phenomena are, simply because the data of our science—observed behavior—do not fully determine the nature of the underlying mechanisms. A further problem is that all of these controversies, phrased as they are in terms of absolute dichotomies, are more likely to be resolved in terms of combined or intermediate explanations rather than either extreme. Debates between those championing environment and those championing heredity are particularly heated because of the societal implications.

There is probably no area of modern psychology in which the question of distinguishing between theories is more immediate than in clinical psychotherapy. Mental illness and human unhappiness are major problems in all human cultures. Unfortunately, there are hundreds, if not thousands, of different approaches, theories, or schools of thought about which strategy represents the most effective means of improving mental health.

The one most salient and compelling observational fact when efforts are made to evaluate the efficacy of psychotherapy is that *all of the methodologies offered by psychotherapy work to some extent and all work equally well*. That is, it really does not matter what strategy is used or what the training of the therapist was. The most extensive studies of the efficacy of psychotherapy were carried out by the Consumer's Union (Anonymous, 1995; Anonymous, 2004). Thousands of patients were asked if their therapy worked, and a large percentage agreed that it had. No particular strategy or theory of therapy, however, seemed to do any better than any other, except for a slight advantage to the behavioral "desensitization" techniques that tended not to delve into the underlying cognitive or psychological mechanisms. More rigorous scientific investigations (Shapiro and Shapiro, 1982; Lipsey and Wilson, 1993) also support this conclusion.

The point is that in the field that is most beset by the greatest variety of theories and explanations of psychological mechanisms and processes, none could be distinguished from any other on the basis of efficacy, of their ability to cure. The psychological "theory" undergirding any one of these psychotherapies was not empirically distinguishable from any other. The conclusion to be drawn for psychotherapy and perhaps all of psychology is that "explanation" is elusive.

5.9 ARE THE METHODS OF PHYSICAL SCIENCE APPROPRIATE FOR PSYCHOLOGY?

Philosophers and theologians have been interested in the mind from the dawn of history. The first scientists and their predecessors—the first natural philosophers, such as Thales of Miletus (624–547), and the greatest, such as Aristotle of Thrace (384–322 BCE)—were as interested in the nature of the mind as that of the physical universe. Their contribution was to make mind, as well as matter, an object of scientific study. From time to time, what we were eventually to call "scientific psychology" has slipped backwards into the domain of philosophy and theology. Nevertheless, since the end of the Renaissance, there has been an acceptance that behavior was worthy of scientific investigation (at least), and the mind could be studied scientifically (at most).

Unfortunately for psychology, the modern scientific method as we now know it first emerged to meet the needs of the physical sciences. It was from Newton's time that a highly mathematical and deductive approach to describing relatively simple mechanisms whose behavior was governed by a few homogenous forces began its remarkable evolution. In the nineteenth century, a few imaginative pioneers, such as Ernst Heinrich Weber (1795–1878), Gustav Theodor Fechner (1801–1887), and Wilhelm Maximilian Wundt (1832–1920), saw the possibility of using the deductive method of the physical sciences to study psychological phenomena. A giant leap was made in setting up the first experimental laboratory for the study of behavior. This was a major strategic change for psychology. What had been restricted to the domain of the armchair now moved into a world of timers and other "brass instruments." Psychologists continue to aspire to utilize the best modern tools, ranging from primitive galvanometers through electronics to the most modern computing and imaging equipment. As Michell (1999) pointed out, when psychologists moved en masse into the laboratory, in the main they bypassed or finessed what should have been a careful consideration of the conceptual foundations of what they were about to do. This was the point at which philsophers were most needed.

Although this approach may have largely been responsible for the transition of psychology from philosophy to the status of an empirical science, some of the initial decisions made by these pioneers had long-term negative effects on the development of psychological science. Laws and methods not suited for psychology were uncritically transferred from physics to psychology. Even some of the first expressions—for example, the derivation of Fechner's law

$$\psi = \ell n \frac{I}{I_0} \qquad \text{Equation 5.1}$$

from Weber's law

$$\frac{\Delta I}{I} = C \qquad \text{Equation 5.2}$$

—were not fully based on the laws of arithmetic. One of the first assumptions made by Fechner was that all of the just noticeable differences (ΔI) equal in perceived size. His proof then proceeded on the basis of this assumption to its conclusion. In a more formal kind of mathematics, such a logical leap would have invalidated the entire proof. Implicit assumptions such as this are ubiquitous through much of psychological theory.

Unfortunately, what this initial physicalization of psychology also did was to suggest methods (e.g., axiomatic-deductive logic and conventional analytic mathematics) that were never to work very well for psychology. Weber's, Fechner's, and all of the other "laws" of behavior that followed have always been considered to be general approximations rather than precision tools comparable to the highly precise laws of physical science.

As the years went by, statistical correlative mathematics eventually took the place in psychology of the kind of analytic mathematics that had been so fruitfully applied to deducing the activity of physical systems. This development suggests that although physical concepts and methods will always play a role in the measurement of behavior, there are evolutionary forces operating in psychology that make the best-suited methodologies different from those appropriate for physics.

As it stands now, most psychological theories are formulated in terms of probability rather than determinacy. Furthermore, despite popular conviction that it is so, there are no instruments that are capable of reading minds, be they animal, human, or computational. There is little, if any, empirical evidence that the polygraph or fMRI can make accessible that which is intrapersonally private. The only means we have, and it is severely limited by what we generally refer to as cognitive penetration, to access a person's mental state is introspection. Sadly, this is known to be wildly inadequate because of incomplete self-knowledge or purposeful deception.

Thus, we must conclude that many of the tools—conceptual, mathematical, instrumental, and classificatory—that have served other sciences so well are not appropriate for the mental sciences.

5.10 CAN MENTAL ENTITIES BE DEFINED?

Efforts to define the entities and objects of our mental life have always been unsuccessful. The basic reason for this failure is that the act of defining something requires reference to something other than the thing itself. Furthermore, the reference term should be as tangible as possible, so that we can concretely, not just metaphorically, relate them to the word being defined. However, references to the private mental world are almost completely devoid of any tangibility or physical reality. The end-product of this is that all definitions of mental events are circular; that is, mental words are defined by reference to terms that have virtually the same meaning as the term being defined. Nothing could make this difficulty of definition clearer than our forlorn efforts to define mind itself. Mind is variously defined as consciousness, awareness, thinking, or the ensemble of processes that go into our mental life, all of which are nothing more than synonyms for the word "mind."

The conclusion we must draw is that mind—a natural process—is indefinable as well as immeasurable. This is a terrible basis on which to build a scientific enterprise. It is necessary to seek out a different strategy, an alternative subject matter, on which to base a true psychological science. The next section of this chapter suggests a new version of an old standby to fill this need.

5.11 TOWARD A SCIENCE OF PSYCHOLOGY

This book is another step in the expression of one critic's view of the limits of mentalist and cognitive psychology. It would be incomplete if I did not propose an alternative approach to achieving a scientific psychology. Of course, it would be wonderful if we could develop a comprehensive, axiomatic-deductive scientific theory of the mind or behavior in the spirit of Newton's Principia. Unfortunately, there are a host of questions that remain unanswered that are not just matters of academic interest, but are also critical to the understanding of our economic, social, and political world. Implicit in the argument I have made here is the conclusion that mentalism and its current embodiment—cognitivism—is not going to be able to confront even the scientific issues, much less those of the applied world. It is unlikely, for example, that we are going to be able to determine how consciousness emerges from the idiosyncratic activity of a huge number of individual neurons.

In this context of the theoretical failure of mentalist cognitivism, what can we suggest for the future of psychology? The answer to this rhetorical question is that we have to look to its past for a strategic approach, one that does not depend on the measurement of the immeasurable or the accessing of the inaccessible; one that links the tangible and ratio-scalable aspects of human behavior to the physical world in which at least some of the tools that have served the physical sciences so well can be applied to increasing our understanding.

A true, as well as useful, science of psychology must temper some of its unrealistic goals and concentrate on what it can do: observe and understand how incident environmental stimuli produce observable responses. This is the intellectual core of the behaviorist approach. Although behaviorism comes in many kinds and flavors, the following scheme presents one version that I think is responsive to the needs and limitations of psychological investigation. A modern behaviorism, and thus a realistic approach to psychological science, must be characterized by the following properties:

1. *Psychophysical*: It must utilize the well-controlled methods of psychophysical research.
2. *Anchored Stimuli*: Stimuli must be anchored to independent physical measures.
3. *Simple Responses*: Psychophysical responses must be limited to simple (Class A, as defined by Brindley, 1960) discriminations such as "same" or "different" to minimize the cognitive penetration effects that distort functional relationships.

4. *Operational*: It must define its concepts in terms of procedures, not in terms of unverifiable, post hoc hypothetical mentalist constructs.
5. *Mathematically Descriptive*: Its formal theories must be acknowledged to be only behaviorally descriptive and to be neutral with regard to underlying mechanisms.
6. *Neuronally Nonreductive*: It must abandon any hope of reducing psychological phenomena to the details of neural nets because of their complexity and the resulting computational intractability.
7. *Experimental*: It must continue to maintain the empirical tussle with nature that has characterized the best psychology in the past.
8. *Molar*: It must look at behavior in terms of the overall unitary integrated process it is and avoid invoking a false modularity.
9. *Empiricist*$_1$ *and Nativist*: It must accept the compromise that both experience and evolved mechanisms motivate and drive behavior.
10. *Empiricist*$_2$ *and Rationalist*: It must accept that compromise that behavior accrues from both stimulus determined (automatic) and logical (inferential) causal sequences.
11. *Anti-Pragmatic*: Psychology must accept its primary role as a theoretical science and base its goals on the quest for knowledge of the nature of our nature, rather than on the immediate needs of society or the utility that some of its findings may seem to have. Useful theories do not necessarily have the same validity as true explanations. This does not negate the profound effect that even the most arcane scientific developments may have on society.

If we follow these principles, psychology can have a fine future and join the rest of modern science in its search for the answers to the great questions that our science, in particular, has the possibility of answering concerning the nature and improvement of the human condition. It is this positive suggestion with which I complete this book.

NOTES

[1]The most common ontological argument of all—"If you do not understand spiritual values, there is no way to convince you"—is probably as true as the equivalent—"If you will not accept evidence, then there is no way to convince you." Both assertions reflect the impenetrable dogmatism of the true believer.

[2]I have paraphrased these conditions into a general language from the strictly biological terminology in which cladistics is usually clothed. Thus, for example, where a biologist might use the word "organism," I use "entities" or "types."

[3]I am fully aware at this point that I am finessing precise definitions of mental terms like thinking, consciousness, and mind. There is a good reason for this omission; precise definitions of these

terms are not easy to come by. Circularity and ambiguity are characteristic of all attempts to define mental terms.

[4]What would interactions between non-sentient automata signify? What would communication between robots accomplish? Could such social interactions drive the evolution of societies and technologies as it did in human societies? Is sentience required for technological and social progress?

[5]One cannot help but call attention once again to Koch's (1992) comments about the "romantic" attachment that psychologists have to physical methods and theories. He stated that "Experimental psychologists have traditionally suffered from a syndrome known as hypermanic physicophilia (with quantificophrenic delusions and methodico-echolalic complications ..." (p.264)

[6]Law 1. Every body perseveres in its state of being at rest or moving uniformly straight forward except insofar as it is compelled to change its state by forces impressed.

Law 2. A change in motion is proportional to the motive force impressed and takes place along the straight line in which that force is impressed.

Law 3. To any action there is always an opposite and equal reaction; in other words, the actions of two bodies on each other are always equal and always opposite in direction. (pp. 416-417)

[7]Once again, I cannot refrain from directing my readers to my earlier book (Uttal, 2005) in which I discussed the possibility that the mind-brain problem might never be solved.

References

Aguirre, A., & Tegmark, M. (2005). Multiple universes, cosmic coincidence, and other dark matters. *Journal of Cosmology and Astroparticle Physics, 0501*, 1-16.

Anderson, J. (1993). *Rules of the Mind.* Mahwah NJ: Erlbaum.

Anderson, R. B. (2001). The power law as an emergent entity. *Memory and Cognition, 29*, 1061-1068.

Anderson, R. B., & Tweney, R. (1997). Artifactual power curves in forgetting. *Memory and Cognition, 25*, 724-730.

Anonymous. (1995). Mental health: Does therapy help? *Consumer Reports, November*, 734-739.

Anonymous. (2000). *Diagnostic and Statistical Manual of Mental Disorders: DSM-IV-TR*: American Psychiatric Association.

Anonymous. (2004). Drugs vs. talk therapy. *Consumer Reports,*October, 22-29.

Anstis, S., & Ramachandran, V. S. (1986). Entrained path deflection in apparent motion. *Vision Research, 26*, 1731-1739.

Bak, P., Tang, C., & Wiesenfeld, K. (1987). Self-organized criticality: An Explanation of 1/f noise. *Physical Review Letters, 59*, 381-384.

Bakan, D. (1954). A generalization of Sidman's results on group and individual functions and a criterion. *Psychological Bulletin, 54*, 63-64.

Barnes, J., G., Howard, R. J., Senior, C., Brammer, M., Bullmore, E. T., Simmons, A., & David, A. S. (1999). The functional effect of the McCollough contingent color after-effect. *Neuroreport, 10*, 195-199.

Bartoshuk, L., Fast, K., Duffy, V. B., Prutkin, J. M., Snyder, D. J., & Green, B. G. (2000). Magnitude matching and a modified LMS produce valid sensory comparisons from PROP studies. *Appetite, 35*, 277 (abstract).

Berlin, B., & Kay, P. (1969). *Basic Color Terms: Their Universality and Evolution*. Berkeley CA: University of California Press.

Biernat, M., & Kobrynowicz, D. (1997). Gender- and race-based standards of competence: Lower minimum standards but higher ability standards from devalued groups. *Journal of Personality and Social Psychology, 72*, 544-557.

Biernat, M., Manis, M., & Kobrynowicz, D. (1997). Simultaneous assimilation and contrast effects in judgments of self and others. *Journal of Personality and Social Psychology, 73*, 254-269.

Biernat, M., Vescio, T. K., & Manis, M. (1998). Judging and behaving toward members of stereotyped groups: A shifting standards perspective. In C. Sedikides & J. Schopler & C. A. Insko (Eds.), *Intergroup Cognition and Intergroup Behavior*. Hillsdale NJ: Erlbaum.

Blanton, H., & Jaccard, J. (2006). Arbitrary metrics in psychology. *American Psychologist, 61*, 27-41.

Blum, H. F. (1968). *Time's Arrow and Evolution*. Princeton NJ: Princeton University Press.

Boyer, C. B. (1989). *A History of Mathematics (2nd Edition). Revised by U. C. Merzbach*. New York: Wiley.

Brady, N., Bex, P. J., & Fredericksen, R. E. (1997). Independent coding across spatial scales in moving fractal images. *Vision Research, 37*, 1873-1883.

Brannon, E. M., & Terrace, H. S. (2000). Representation of the numerosities by rhesus macaques (Macaca mulatta). *Journal of Experimental Psychology: Animal Behavior Processes, 26*, 31-49.

Brannon, E. M., & Walle, V. D. (2001). The development of ordinal numerical competence in young children. *Cognitive Psychology, 43*, 53-81.

Brindley, G. S. (1960). *Physiology of the Retina and Visual Pathway*. London: Edward Arnold.

Brown, J. F., & Voth, A. C. (1937). The path of seen motion as a function of the vector-field. *American Journal of Psychology, 49*,. 543-563.

Burns, E. M., & Ward, W. D. (1978). Categorical perception - phenomenon or epiphenomenon: Evidence from experiments in the perception of melodic musical intervals. *Journal of the Acoustical Society of America, 63*, 456-468.

Busby, T. A., & Townsend, J. T. (2001). Independent sampling vs. inter-item dependencies in whole report processing: Contributions of processing architecture and variable attention. *Journal of Mathematical Psychology, 45*, 283-323.

Butz, W. P., & Torrey, B. B. (2006). Some frontiers in social science. *Science, 312*, 1898-1900.

Caelli, T., Hoffman, W., & Lindman, H. (1978). Apparent motion:Self-excited oscillation induced by retarded neuronal flows. In E. L. J. Leeuwenberg & H. F. J. M. Buffart (Eds.), *Formal Theories of Visual Perception*. New York: Wiley.

Carpenter, S. (2000). A taste experts sniffs out a long-standing measurment oversight. *Monitor on Psychology, 31*.

Chan, K. S., Kristofferson, A. B., & Stenseth, N. C. (2003). Burmann expansion and test for additivity. *Biometrika, 90*, 209-222.

Cheng, P. W. (1997). From covariation to causation: A causal power theory. *Psychological Review, 104*, 367-405.

Clayton, K., & Frey, B. (1995, August 1995). *Studies of mental noise.* Paper presented at the Fifth Annual Conference of the Society for Chaos Theory in Psychology and the Life Sciences., Garden City NY.

Coltheart, M. (2006a). What has functional neuroimaging told us about the mind (so far)? *Cortex, 42*, 323-331.

Coltheart, M. (2006b). Perhaps functional neuroimaging has not told us anything about the mind (so far). *Cortex, 42*, 422-427.

Cook, N. D., Hayashi, T., Amemiya, T., Suzuki, K., & Leuhman, L. (2002). Effects of visual-field inversions on the reverse-perspective illusion. *Perception, 31*, 1147-1151.

Coombs, C. H. (1950). Psychological scaling without a unit of measurement. *Psychological Review, 57*, 145-158.

Coren, S., Girgus, J. S., Erlichman, H., & Hakstian, A. R. (1976). An empirical taxonomy of visual illusions. *Perception and Psychophysics, 20*, 129-137.

Dawes, R. M., & Smith, T. L. (1985). Attitude and opinion measurement. In G. Lindzey & E. Aronson (Eds.), *The Handbook of Social Psychology* (pp. 509-566). New York: Random House.

Day, R. H. (1972). Visual spatial illusions: A general explanation. *Science, 175*, 1335- 1340.

Dehaen, S., Izard, V., Pica, P., & Spelke, E. (2006). Core knowledge of geometry in an Amazonian indigene group. *Science, 311*, 381-384.

Deutsch, D. Henthorn, T., & Dolson, M. (2004). Absolute pitch, speech, and tone language: Some experiments and a proposed framework. *Music Perception, 21*, 339-356.

Deutsch, D., Henthorn, T., Marvin, E., Xu, H-S. (2006). Absolute pitch among American and Chinese conservatory students: Prevalence difference, and evidence for a speech-related critical period (L). *Journal of the Acoustical Society of America, 119*, 719-722.

DeValois, R. L., Smith, C. J., Kitai, S. T., & Karoly, A. J. (1958). Response of single cells in monkey lateral geniculate nucleus to monochromatic light. *Science, 127*, 238-239.

DiLollo, V. (1980). Temporal integration in visual memory. *Journal of Experimental Psychology: General, 109*, 75-97.

Donders, F. C. (1868/1969). On the speed of mental processes (Translated by W. G. Koster). *Acta Psychologica, 30*, 412-431.

Dreyer, G. (1998). *Umm el Qaab I: Das predynastiche Konigsgrab U-j und seine fruhen Schriftzeugnisse.* Mainz: Verlag Philipp von Zabern.

Dreyer, G. (1999). Abydos: Umm el-Qa'ab. In B. K. A. (Ed.), *Encyclopedia of the Archeology of Ancient Egypt* (pp. 109-114). London: Routledge.

Eagleman, D. M., & Sejnowski, T. J. (2000). Motion integration and postdiction in visual awareness. *Science, 287*, 2036-2038.

Eagleman, D. M., & Sejnowski, T. J. (2003). The line-motion illusion can be reversed by motion signals after the line disappears. *Perception, 32*, 963-968.

Einstein, A. (1905). On the electrodynamics of moving bodies (Zur Elektrodynamik bewegter Körper). *Annalen der Physik., 17*, 891-921.

Einstein, A. (1917/2003). *Relativity: The Special and General Theory: Translated by Lawson, R. W. and Republished in Electronic Form*
(http://www.marxists.org/reference/archive/einstein/works/1910s/relative/).
New York: Henry Holt and Company.

Eriksen, C. W., & Collins, J. F. (1965). Reinterpretation of one form of backward and forward masking in visual perception. *Journal of Experimental Psychology, 70*, 343-351.

Estes, W. K. (1956). The problem of inference from curves based on group data. *Psychological Bulletin, 53*, 134-140.

Estes, W. K. (2002). Traps in the route to models of memory and decision. *Psychonomic Bulletin and Review, 9*, 3-25.

Etcoff, N. L., & Magee, J. J. (1992). Categorical perception of facial expressions. *Cognition, 44*, 227-240.

Everett, D. L. (2005). Cultural constraints on grammar and cognition in Piraha. *Current Anthropology, 46*, 621-646.

Exner, S. (1875). Experimentelle Untersuchung der einfachisten psychischen Processe III. *Pflugers Arkiv fur die gesammte Physiolgie des Menschen und Thiere, 11*, 402-412.

Falmagne, J.-C. (2004). Meaningfulness and order-invariance: Two fundamental principles for scientific laws. *Foundations of Physics, 34*, 1341-1384.

Fast , K., Green, B. G., Snyder, D. J., & Bartoshuk, L. (2001). Remembered intensities of taste and oral burn correlate with PROP bitterness. *Chemical Senses*.

Fechner, G. (1860/1966). *Elements of Psychophysics* (H. E. Adler, Trans.). New York: Holt, Rinehart, and Winston.

Fisher, G. H. (1973). Towards a new explanation for the geometrical illusions. *British Journal of Psychology, 64*, 607-621.

Fraisse, P. (1963). *The Psychology of Time*. New York: Harper and Row.

Friedman, W. J. (1990). *About time: Inventing the fourth dimension*. Cambridge MA: MIT Press.

Garner, W. R. (1953). An information analysis of absolute judgments of loudness. *Journal of Experimental Psychology, 46*, 373-380.

Gellert, W., Kustner, H., Hellwich, M., & Kastner, H. (Eds.). (1977). *The VNR Concise Encyclopedia of Mathematics*. New York: Van Nostrand Reinhold.

Gibb, R. W. (2007). Visual spatial disorientation. Revisiting the black hole illusion. Aviation, Space, and Environmental Medicine, 78, 801-888.

Gilden, D. L. (1997). Fluctuations in time required for elementary decisions. *Psychological Science, 8*, 296-301.

Gilden, D. L. (2001). Cognitive emissions of 1/f noise. *Psychological Review, 108*, 33-56.

Gilden, D. L., Thornton, T., & Mallon, M. W. (1995). 1/f noise in human cognition. *Science, 267*, 1837-1839.

Gogel, W. C. (1990). A theory of phenomenal geometry and its applications. *Perception and Psychophysics, 48*, 105-123.

Gordon, P. (2004). Numerical cognition without words: Evidence from Amazonia. *Science, 306*, 496-499.

Green, D. M., & Luce, R. D. (1974). Variability of magnitude estimates: A timing theory analysis. *Perception and Psychophysics, 15*, 291-300.

Gregory, R. L. (1963). Distortion of visual space as inappropriate constancy scaling. *Nature, 199*, 678-680.

Grimes, J. (1996). On the failure to detect changes in scenes across saccades. In K. Akins (Ed.), *Perception* (Vol. 89-110). New York: Oxford University Press.

Hake, H. W., & Garner, W. R. (1951). The effect of presenting various numbers of discrete steps on scale reading accuracy. *Journal of Experimental Psychology, 42*, 358-366.

Hall, D. (2006). *Elementary Statistical Methods I*. Indiana University Purdue University Indianapolis Department of Mathematical Sciences. Retrieved April 24, 2006, from the World Wide Web: www.math.iupui.edu/handouts/ch.1

Hamlin, A. J. (1895). On the least observable interval between stimuli addressed to disparate senses and to different organs of the same sense. *American Journal of Psychology, 6*, 564-573.

Hand, D. J. (2004). *Measurement: Theory and Practice*. London: Arnold.

Harnad, S. R. (1987). *Categorical Perception: The Groundwork of Cognition*. Cambridge: Cambridge University Press.

Hausdorff, J. M., & Peng, C.-K. (1996). Multi-scaled randomness: A source of 1/f noise in biology. *Physical Review E, 54*, 2154-2157.

Hauser, M. D., Carey, S., & Hauser, L. B. (2000). Spontaneous number representations in wild rhesus monkeys. *Proceedings of the Royal Academy of Sciences, 93*, 1514-1517.

Heathcote, A., Brown, S., & Mewhort, D. J. (2000). The power law repealed: The case for an exponential law of practice. *Psychonomic Bulletin and Review, 7*, 185-207.

Heisenberg, W. (1927). Uber den anschaulichen Inhalt der quantentheoretischen Kinematic und Mechanik. *Z. fur Phys, 43*, 172-198.

Helmholtz, H. v. (1856). *Handbuch der Physiologishen Optik*. Leipzig: Voss.

Hennig, W. (1966). *Phylogenetic Systematics* (D. D. Davis & R. Zangerl, Trans.). Urbana: University of Illinois Press.

Hick, W. E. (1952). On the rate of gain of information. *Quarterly Journal of Experimental Psychology, 4*, 11-26.

Hilgetag, C. C., O'Neil, M. A., & Young, M. P. (1996). Indeterminate organization of the visual system. *Science, 271*, 776-777.

Hilgetag, C. C., O'Neil, M. A., & Young, M. P. (2000). Hierarchical organization of macaque and cat cortical sensory mechanisms explored with a novel network processor. *Philosophical Transactions of the Royal Society of London. B., 355*, 71-89.

Hoffman, W. C. The Lie algebra of visual perception. *Journal of Mathematical Psychology, 3*, 65-98.

Holder, O. (1901). Die Axiome der Quanitat die Lehre vom Mass. *Berichte uber die Verhandlungen der Koniglich Sachsischen Gesellschaft der Wissenschaften Leipzig, Mathematisch-Physische Klasse, 53*, 1-46.

Howe, C. Q., & Purves, D. (2005). The Poggendorf illusion explained by natural scene geometry. *Proceedings of the National Academy of Sciences, 102*, 7707-7712.

Howe, C. Q., & Purves, D. (2005a). Natural-scene geometry predicts the perception of angles and line orientation. *Proceedings of the National Academy of Sciences, 102*, 1228-1233.

Howe, C. Q., & Purves, D. (2005b). The Muller-Lyer illusion explained by the statistics of image-source relationships. *Proceedings of the National Academy of Sciences, 102*, 1234-1239.

Howe, C. Q., Yang, Z.& Purves, D. (2005). The Poggendorf Illusion explained by natural scene geometry. *Proceedings of the National Academy of Sciences, 102*, 7707-7712.

Huk, A. C., Ress, D., & Heeger, D. J. (2001). Neuronal basis of the motion aftereffect reconsidered. *Neuron, 32*, 161-172.

Hull, C. L. (1943). *Principles of Behavior: An Introduction to Behavior Theory*. New York: D. Appleton-Century Company.

Hull, C. L. (1952). *A Behavior System: An Introduction to Behavior Theory*. New Haven: Yale University Press.

Humphrey, G. K., & Goodale, M. A. (1998). Probing unconscious visual processing with the McCollough effect. *Consciousness and cognition, 7*, 494-519.

Hyman, R. (1953). Stimulus information as a deterrminant of reaction time. *Journal of Experimental Psychology, 45*, 423-432.

Isaacson, W. (2007). *Einstein: His Life and Universe*. New York: Simon and Schuster.

Jevons, W. S. (1871). The power of numerical discrimination. *Nature, 3*, 282-282.

Jones, D. D., & Holding, D. H. (1975). Extremely long-term persistence of the McCollough effect. *Journal of Experimental Psychology: Human Perception and Performance, 1*, 323-327.

Jones, F. N., & Marcus, M. J. (1961). The subject effects in judgments of subjective magnitude. *Journal of Experimental Psychology, 61*, 40-44.

Judd, D. B. (1932). Chromaticity sensitivity to stimulus differences. *Journal of the Optical Society of America, 22*, 72-108.

Kahneman, D. (1968). Metacontrast: Method, findings, and theory in studies of visual masking. *Psychological Bulletin, 70*, 404-425.

Karabatsos, G. (2005). Additivity tests. In B. S. Everitt & D. C. Howell (Eds.), *Encyclopedia of Statistics in Behavioral Science*. Chichester: John Wiley and Sons.

Kaufman, E. L. (1974). *An Introduction to Perception*. New York: Oxford.

Kaufman, E. L., Lord, M. L., Reese, T. W., & Volkman, J. (1949). The discrimination of visual numbers. *American Journal of Psychology, 62*, 498-525.

Kawabata, N., Yamagami, K., & Nokai, M. (1978). Visual fixation points and depth perception. *Visual Research, 18*, 853-854.

Kay, P., & Regier, T. (2003). Resolving the question of color naming universals. *Proceedings of the National Academy of Sciences, 100*, 9085-9089.

Kendall, M. G., & Babington Smith, B. (1938). Randomness and random sampling numbers. *Journal of the Royal Statistical Society, 101*, 147-166.

Killeen, P. (1994). Mathematical principles of reinforcement. *Behavioral and Brain Sciences, 17*, 105-172.

Klemmer, E. T., & Frick, F. C. (1953). Assimilation of information from dot and matrix patterns. *Journal of Experimental Psychology, 45*, 15-19.

Koch, S. (1954). Clark L. Hull. In A. T. Poffenberger (Ed.), *Modern Learning Theory* (pp. 1-176). New York: Appleton-Century-Crofts.

Koch, S. (1992). Psychology's Bridgman vs. Bridgman's Bridgman. *Theory and Psychology, 2*, 261-290.

Koenderink, J. J., van Dorn, J., Kappers, A. M. & Todd, J.T. Pointing out the picture. *Perception*, 33, 513-530.

Kolers, P. A. (1972). *Aspects of Motion Perception*. Oxford: Pergamon Press.

Korte, A. (1915). Kinematoskopische untersuchungen. *Zeitschrift fur Psychologie, 72*, 193-296.

Levin, D. T., & Simons, D. J. (1997). Failure to detect changes in attended objects in motion pictures. *Psychonomic Bulletin and Review, 4*, 501-506.

Levitin, D. J., & Rogers, S. E. (2005). Absolute pitch: Perception, coding, and controversies. *Trends in Cognitive Sciences, 9*, 26-33.

Liberman, A.M., Harris, K. S., Hoffman, H. S., & Griffith, B. S. (1957). The discrimination of speech sounds within and across phoneme boundaries. *Journal of Experimental Psychology, 61*, 379-388.

Lipsey, M. A., & Wilson, D. B. (1993). The efficacy of psychological, educational, and behavioral treatment:Confirmation from meta-analysis. *American Psychologist, 48*, 1181-1209.

Luce, R. D. (1995). Why should mathematics play a role in psychology? *Annual Review of Psychology, 46*, 1-26.

MacCorquodale, K., & Meehl, P. E. (1948). On a distinction between hypothetical constructs and intervening variables. *Psychological Review, 55*, 95-107.

MacNichol, E. J., & Svaetician, G. (1958). Electric responses of isolated retinas of fishes. *American Journal of Ophthalmology, 46*, 26-46.

Mandelbrot, B. B. (1961). On the theory of word frequencies and on related Markovian models of discourse. In R. Jacobson (Ed.). (pp. 190-219), *Structures of Language and Its Mathematical Aspects*. New York: American Mathematical Society.

Mandelbrot, B. B. (1999). *Multifractals and 1/f Noise: Wild Self-Affinity in Physics (1963-1976)*. New York: Springer-Verlag.

Marchal, C. (2005). *Henri Poincare':A decisive contribution to relativity*. Office National d'Etudes et de Recherches Aerospatiales. Retrieved, from the World Wide Web: http://www-cosmosaf.iap.fr/poincare.htm

Massaro, D. W. (1997). *Perceiving Talking Faces: From Speech Perception to a Behavioral Principle*. Cambridge MA: MIT Press.

McBeath, M. K. (1990). The rising fastball: Baseball's impossible pitch. *Perception, 19*, 545-552.

McBeath, M. K., & Shepard, R. N. (1989). Apparent motion between shapes differing in location and orientation: A window technique for estimating path curvature. *Perception and Psychophysics, 46*, 333-337.

McCollough, C. (1965). Color adaptation of edge-detectors in the human visual system. *Science, 149*, 1115-1116.

McConkie, G. W., & Currie, C. B. (1996). Visual stability across saccades while viewing complex pictures. *Journal of Experimental Psychology: Human Perception and Performance, 22*, 563-581.

McKeon, C. R. (Ed.). (1941). *The Basic Works of Aristotle*. New York: Random House.

Meissner, C. A., & Brigham, J. C. (2001). Thirty years of investigating the own-race bias in memory for faces. A meta-analytic review. *Psychology, Public Policy, and the Law, 7*, 3-35.

Messick, S. (1995). Validity of psychological measurement: Validation of inferences from person's responses and performances into score meaning. *American Psychologist, 50*, 741-749.

Meyer, M. L., & Shipley, T. F. (2003). Perception of curved apparent motion paths [Abstract]. *Journal of Vision, 3*, 785a.

Michell, J. (1997). Quantitative science and the definition of measurement in psychology. *British Journal of Psychology, 88*, 355-383.

Michell, J. (1999). *Measurement in Psychology: Critical History of a Methodological Concept*. Cambridge, England: Cambridge University Press.

Miller, G. A. (1956). The magical number seven, plus or minus two: Some limits on our capacity for processing information. *Psychological Review, 63*, 81-97.

Minkel, J. R. (2006). Hard landscape: Finding our universe in string theory appears impossible. *Scientific American, 294*(6), 20-21.

Minkowski, H. (1908/1952). Space and time. New York: Dover, pp. 75-91. In H. A. Lorentz, A. Einstein, H. Minkowski & H. Weyl (Eds.), *The Principle of Relativity: A Collection of Original Memoirs on the Special and General Theory of Relativity*. New York: Dover.

Mix, K., Huttenlocher, J., & Levine, S. C. (1996). Do preschool children recognize auditory-visual correspondences? *Child Development, 67*, 1592-1608.

Mollon, J. D., & Perkins, A. J. (1996). Errors of judgment at Greenwich in 1796. *Nature, 380*, 101-102.

Moore, E. F. (1956). Gedanken-experimental studies on experimental machines. In C. E. Shannon & J. McCarthy (Eds.), *Automata Studies*. Princeton, NJ: Princeton University Press.

Mulcahy, N. J., & Call, J. (2006). Apes save tools for future use. *Science, 312*, 1038-1040.

Narens, L. (1996). A theory of magnitude estimation. *Journal of Mathematical Psychology, 40*, 109-129.

Necker, L. A. (1832). On an apparent change of position in a drawing of engraved figure of a crystal. *Philosophical Magazine, 1*.

Neumann, O. Niepel, M. (2004). Timing of "perception" and perception of "time". In C. Kaernbach & E. Schroeger & H. Muller (Eds.), *Psychophysics Beyond Sensation*. Mahwah NJ: Erlbaum.

Newell, A. (1990). *Unified Theory of Cognition*. Cambridge MA: Harvard University Press.

Newton, I. (1687). *Philosophiae Naturalis Principia Mathematica (Mathematical Principles of Natural Philosophy)*. London: S. Pepys, Reg. Soc.,PRAESES.

Nimh, S. D. (1976). Polynomial law of sensation. *American Psychologist, 31*, 808-809.

Pachella, R. G. (1974). The interpretation of reaction time in information processing research. In B. H. Kantowitz (Ed.), *Human Information Processing:Tutorials in Performance and Cognition* (pp. 41-82). Hillsdale NJ: Erlbaum.

Pearson, P. M., & Schaefer, E. G. (2005). Toupee of not toupee? The role of instructional set, centrality, and relevance in change blindness. *Visual Cognition, 12*, 1528-1543.

Penfield, W. (1955). The permanent record of the stream of consciousness. *Acta Psychologica, 11*, 47-69.

Penzias, A. A. & Wilson, R.W. (1965) A measurement of excess antenna temperature at 4080 Mc. *The Astrophysical Journal, 142*, 419

Piaget, J. (1952). *The Childs Conception of Number*. New York: Norton.

Piazza, M., A., M., Butterworth, B., & Price, C. J. (2002). Are subitizing and counting implemented as separate or functionally overlapping processes? *Neuroimage, 15*, 435-446.

Pica, P., Lemer, C., Izard, V., & Dehaene, S. (2004). Exact and approximate arithmetic in an Amazonian indigene group. *Science, 306*, 499-503.

Poincaré, H. (1902). *La Science et l'Hypothese*. Paris: Flammarion.

Poincaré, H. (1904). L'état actuel et l'avenir de la physique mathématique. *Bulletin des Sciences Mathématiques, 28*, 302-324.
Pollack, I. (1952). The information of elementary auditory displays. *Journal of the Acoustical Society of America, 24*, 745-749.
Pollack, I. (1953). The information of elementary auditory displays II. *Journal of the Acoustical Society of America, 25*, 765-769.
Pressing, J. (1999). *Sources of 1/f noise effects in human cognition and performance.* Retrieved, from the World Wide Web: http://www.paideusis.matco.ro/n1ciss.html#jp
Pressing, J., & Jolley-Rogers, G. (1997). Spectral properties of human cognition and skill. *Biological Cybernetics, 76*, 339-347.
Purves, D., Lotto, R. B., William, S. M., Nundy, S., & Yang, Z. (2001). Why do we see things the way we do? Evidence for a wholly empirical strategy of vision. *Philosophical Transactions of the Royal Society of London B, 356*, 285-297.
Rashevsky, N. (1948). *Mathematical Biophysics.* Chicago: University of Chicago Press.
Reichenbach, H. (1956). *The Direction of Time.* Berkeley CA: University of California Press.
Resink, R. A. (2002). Change detection. *Annual Review of Psychology, 53*, 245-277.
Resink, R. A., O'Regan, J. K., & Clark, J. (1997). To see or not to see: The need for attention to perceive changes in scenes. *Psychological Science, 8*, 368-373.
Riggs, L. A., White, K. D., & Eimas, P. D. (1974). Establishment and decay of orientation-contingent aftereffects of color. *Perception and Psychophysics, 16*, 535-542.
Roberts, S., & Pashler, L. (2000). How persuasive is a good fit? A comment on theory testing. *Psychological Review., 107*, 358-367.
Robinson, J. O. (1972). *The Psychology of Visual Illusion.* London: Hutchinson University Library.
Rubin, E. (1921). *Visuelle wahrgenommene figurer.* Copenhagen: Gyldendalske.
Sachs, R. G. (1987). *The Physics of Time Reversal.* Chicago: University of Chicago Press.
Sacks, O. (2004). In the river of consciousness. *New York Review of Books, 51.*
Schroeder, M. (1991). *Fractals, Chaos, and Power Laws. Minutes from an Infinite Paradise.* New York: Freeman.
Schutter, D. J. L. G., de Haan, E. H. F., & van Honk, J. (2006). Scaling problems in the brain-mind continuum. *Cortex, 42*, 411-413.
Shallice, T. (1964). The detection of change and the perceptual moment hypothesis. *British Journal of Statistical Psychology, 17*, 113-135.
Shapiro, D., & Shapiro, D. (1982). Meta-analysis of comparative therapy outcome studies: A replication and refinement. *Psychological Bulletin, 92*, 581-604.
Shepard, R. N., & Zare, S. L. (1983). Path-guided apparent motion. *Science, 220*, 632-634.
Shiffrar, M., & Freyd, J. J. (1990). Apparent motion of the human body. *Psychological Science, 1*, 257-264.
Shower, E. G., & Biddulph, R. (1931). Differential pitch sensitivity of the ear. *Journal of the Acoustical Society of America, 3*, 275-287.
Sidman, M. (1952). A note on the functional relations obtained from group data. *Psychological Bulletin, 49*, 263-269.
Siegel, S. (1957). Nonparametric statistics. *American Statistician, 11*, 13-19.

Simon, T. J. (1999). The foundations of numerical thinking in a brain without numbers. *Trends in Cognitive Science, 3*, 363-365.

Simons, D. J., & Ambinder, M. S. (2005). Change blindness: Theory and Consequences. *Current Trends in Psychological Science, 14*, 44-48.

Simons, D. J., & Chabris, C. F. (1999). Gorillas in our midst: Sustained inattention blindness for dynamic events. *Perception, 28*, 1059-1074.

Simons, D. J., & Levin, D. T. (1998). Failure to detect changes to people in a real world interaction. *Psychonomic Bulletin and Review, 5*, 644-649.

Smith, W. F. (1933). The relative quickness of visual and auditory perception. *Journal of Experimental Psychology, 16*, 239-270.

Smolin, L. (2006). *The Trouble with Physics: The Rise of String Theory, the Fall of a Science, and What Comes Next*. New York: Houghton Mifflin.

Smythies, J. (2003). Space, time, and consciousness. *Journal of Consciousness Studies, 10*, 47-56.

Snyder, A. W., & Mitchell, D. J. (1999). Is integer arithmetic fundamental to mental processing?: The mind's secret arithmetic. *Proceedings of the Royal Society of London (B), 266*, 587-592.

Spelke, E., & Dehaen, S. (1999). Biological foundations of numerical thinking: Response to T. J. Simon. *Trends in Cognitive Science, 3*, 365-366.

Spiegel, A. (2005). The dictionary of disorder: How one man revolutionized psychiatry. *New Yorker, January 3, 2005*, 56-63.

Spillmann, L. (1999). From elements to perception: Local and global processing in visual neurons. *Perception, 28*, 1461-1492.

Sternberg, S. (1969). The discovery of processing stages: Extension of Donder's method. *Acta Psychologica, 30*, 276-315.

Stevens, S. S. (1946). On the theory of scales of measurement. *Science, 103*, 677-680.

Stevens, S. S. (1951). Mathematics, Measurement, and Psychophysics. In S. S. Stevens (Ed.), *Handbook of Experimental Psychology*. New York: John Wiley and Sons.

Stevens, S. S. (1971a). Neural events and the psychophysical law. *Science, 170*, 1043-1050.

Stevens, S. S. (1971b). Issues in psychophysical measurement. *Psychological Review, 78*, 426-450.

Stevens, S. S. (1975). *Psychophysics*. New York: Wiley.

Stewart, A. L., & Pinkham, R. S. (1991). A space-variant operator for visual sensitivity. *Biological Cybernetics, 64*, 373-379.

Stewart, A. L., & Pinkham, R. S. (1994). Space-variant models of visual acuity using self-adjoint integral operators. *Bilogical Cybernetics, 71*, 161-167.

Stigler, R. (1910). Chronophotische studien uber den umgebungs kontrast. *Pflugers Arkiv fur die gesammte Physiolgie des Menschen und Thiere, 134*, 365-435.

Stigler, S. M. (1999). *Statistics on the Table: The History of Statistical Concepts and Methods*. Cambridge MA: Harvard University Press.

Stroud, J. M. (1949). The psychological moment in perception. In H. von Foerster (Ed.), *Transactions of the Sixth Conference on Cybernetics*. New York: Josiah Macy, Jr.

Suppes, P., & Zinnes, J. L. (1963). Basic measurement theory. In R. D. Luce,R. R. Bush & E. Galanter (Eds.), *Handbook of Mathematical Psychology, Volume 1*. New York: John Wiley and Sons.

Treffert, D. A., & Christensen, D. D. (2005). Inside the mind of a savant. *Scientific American, 293*, 108-113.

Tufte, E. R. (1983). *The Visual Display of Quantitative Information*. Cheshire CN: Graphics Press.

Tukey, J. W. (1949). One degree of freedom for non additivity. *Biometrics, 5*, 232-242.

Turchin, P. (2006). *War & Peace & War: The Life Cycles of Imperial Nations*. New York: Pi Press.

Uttal, W. R. (1969). Masking of alphabetic character recognition by dynamic visual noise. *Perception and Psychophysics, 6*, 121-127.

Uttal, W. R. (1973). *The Psychobiology of Sensory Coding*. New York: Harper and Row.

Uttal, W. R. (1981). *A Taxonomy of Visual Processes*. Hillsdale NJ: Erlbaum.

Uttal, W. R. (1998). *Toward a New Behaviorism: The Case against Perceptual Reductionism*. Mahwah NJ: Erlbaum.

Uttal, W. R. (2001). *The New Phrenology: The Limits of Localizing Cognitive Processes in the Brain*. Cambridge MA: MIT Press.

Uttal, W. R. (2003). *Psychomythics: Sources of Artifacts and Misconceptions in Scientific Psychology*. Mahwah NJ: Erlbaum.

Uttal, W. R. (2005). *Neural Theories of Mind: Why the Mind-Brain Problem May Never Be Solved*. Mahwah NJ: Erlbaum.

Uttal, W. R. (2007). *The Immeasurable Mind*. Amherst NY: Prometheus Books.

Vandermeer, J., & Perfector, I. (2006). A keystone mutualism drives pattern in a power function. *Science, 311*, 1000-1002.

Van Orden, G. C., Moreno, M. A., & Holden, J. G. (2003). A proper metaphysics for cognitive performance. *Nonlinear Dynamics, Psychology and Life Sciences, 7*, 49-60.

VanRullen, R., & Koch, C. (2003). Is perception discrete or continuous. *Trends in Cognitive Science, 7*, 207-213.

Wagenmakers, E.-J., Farrell, S., & Ratcliff, R. (2004). Estimation and interpretation of 1/f noise in human cognition. *Psychonomic Bulletin and Review, 11*, 579-615.

Ward, W. D. (1963). Absolute pitch, Part II. *Sound, 2*, 33-41.

Ward, W. D. (1999). Absolute pitch. In D. Deutsch (Ed.), *The Psychology of Music* (pp. 265-298). San Diego: Academic Press.

Watson, A. B. (1978). A Riemann geometric explanation of the visual illusions and figural aftereffects. In E. L. J. Leeuwenberg & H. F. J. M. Buffart (Eds.), *Formal Theories of Visual Perception*. New York: Wiley.

Wegner, D. M., Schneider, D. J., Carter III, S. R., & White, T. L. (1987). Paradoxical effects of thought suppression. *Journal of Personality and Social Psychology, 53*, 5-13.

Weisstein, N. (1972). Metacontrast. In D. Jameson & Hurvich, L. M. (Eds.), *Handbook of Sensory Psychology: Visual Psychophysics (Vol. VII/4)*. New York: Springer Verlag.

Wertheimer, M. (1912). Experimentelle studien uber das sehen von bewegung. *Zeitschrift fur Psychologie, 61*, 161-265.

Whitehead, A. N., & Russell, B. (1910-1913). *Principia Mathematica*. Cambridge: Cambridge University Press.

Wigner, E. (1960). The unreasonable effectiveness of mathematics in the natural sciences. *Communications in Pure and Applied Mathematics,, 13*, 1-13.

Woit, P. (2006). *Not Even Wrong: The Failure of String Theory and the Search for Unity in Physical Law*. New York: Basic Books.

Woodrow, H. (1934). The temporal indifference interval determined by the method of mean error. *Journal of Experimental Psychology, 17*, 167-188.

Wynn, K. (1990). Children's understanding of counting. *Cognition, 36*, 155-193.

Yule, G. U. (1926). Why do we sometimes get nonsense-correlations between time series? A study in sampling and the nature of time series. *Journal of the Royal Statistical Society, 89*, 1-64.

Zanuttini. (1976). A new explanation for the Poggendorf illusion. *Perception and Psychophysics, 20*, 29-32.

Zipf, G. K. (1935/1965). *Psycho-biology of Languages*. Cambridge MA: MIT Press.

Name Index

Subject Index

?rinted in the USA/Agawam, MA
)ctober 25, 2014
599583.076